THE EARTH GENERATED
AND ANATOMIZED

THE EARTH GENERATED
AND ANATOMIZED
by William Hobbs

*An early eighteenth century
theory of the earth*

Edited with an introduction by

ROY PORTER

BRITISH MUSEUM (NATURAL HISTORY)

CORNELL UNIVERSITY PRESS
Ithaca and London

©Trustees of the British Museum (Natural History) 1981

First published 1981 by Cornell University Press.

Library of Congress Cataloging in Publication Data

Hobbs, William.
The earth generated and anatomized.

Includes bibliographical references.
1. Earth sciences—Early works to 1800.
I. Porter, Roy, 1946-   II. Title
QE25.H62   551   80-67632
ISBN 0-8014-1366-4

PRINTED IN THE UNITED STATES OF AMERICA

Contents

Introduction

The discovery of new scientific manuscripts is always exciting: the discovery of a new scientific author even more so. Hence the appearance in 1973 of a new early eighteenth century theory of the Earth, *The earth generated and anatomized*, by an unknown writer, William Hobbs, was particularly significant, for many of the author's discussions of the problems of the terraqueous globe are sufficiently skilful in themselves to merit his theory being regarded as one of the more constructive attempts of its day to interpret the Earth. Hobbs's theory is also of significance in that it draws deeply upon long-established traditions of natural philosophical discourse, such as alchemy and organic theories of Nature, to offer a conscious challenge to the fashionable mechanical philosophy and the Newtonian world-picture. Furthermore, reading between its lines tells us much about the conditions under which amateurs, across provincial England, were beginning to develop a taste for scientific inquiry at the beginning of the eighteenth century. Scholars should hence be grateful to the British Museum (Natural History) for purchasing this important manuscript and undertaking to produce a printed version of it for the first time.

I am only too conscious of the imperfect and provisional state of much of the material in my commentary and references. To some degree it is because William Hobbs seems to have been a genuinely obscure and shadowy figure. Considerable research in a large number of archives over the last three years has failed to turn up more than a handful of references to him (I am still not even totally sure as to the identity of the author, given that there were at least two William Hobbs—father and son—living in Weymouth at the time the treatise was written!). I have thought it better to make the text available in print at an early date, in order that Hobbs's treatise could actually be read by a wider audience, rather than pause for perhaps many further years in the hope of tracking down additional biographical material for a definitive edition. I hope that the appearance of this edition in print will actually lead to the uncovering of more information about its absorbing author. In producing this edition I have built upon a preliminary paper which I delivered to a conference of the Society for the Bibliography of Natural History held in Spring, 1975, and which was published, as 'William Hobbs of Weymouth and his *The earth generated and anatomized* (?1715)' in the *Journal of the Society for the Bibliography of Natural History* 7 1976: 333–41; I have here corrected some errors in that account, dropped some hypotheses and verified others.

The kindness, patience and knowledge of innumerable people have been imposed upon in the course of my work on this manuscript—far more than I can mention here. I should like especially to thank the staff of the Department of Palaeontology at the British Museum (Natural History) for their unfailing helpfulness: Anthony P. Harvey in particular. The staffs of many libraries, record offices, museums and repositories have helped me track down materials and have answered numerous letters of inquiry. I should like particularly to mention the Dorset County Record Office, the Dorset Natural History and Archaeological Society, The Royal Society, The British Library (especially Dr C. Wright), The Custom House, Weymouth, and H.M. Customs and Excise.

Amongst the many individuals who have helped me and supplied me with information I should like to single out Miss M. Weinstock, Mr H. West and Mr V. Adams for their knowledge of Dorset; Dr H. Torrens for his unfailing geological help; Mr Peter Croft, Librarian of King's College, Cambridge for his advice on the calligraphy of the Manuscript; Dr S. Conway Morris and my wife, Sue Porter, for their great help in reconstructing Hobbs's diagrams; Miss M. Deacon for patiently instructing me in many aspects of the science of the sea; and Mr D. Bryden, Dr W. E. Knowles Middleton, Mr S. Schaffer, Dr J. Schuster, Professor P. Grierson, Dr K. Figlio, Professor A. R. Hall, Mr. A. Turner, Mr G. L'E. Turner, Dr C. Webster, Dr Roger French for helping me to hunt down references. Mr J. B. Morrell, Mr A. P. Harvey and Mr S. Schaffer kindly read earlier drafts of the 'Introduction' and offered helpful criticisms. Miss Fiona Mainstone undertook some search work, and Miss Katy Henkel admirably transcribed and typed the text. I must alone be responsible for the errors and gaps which still exist.

I should like to thank the following for permission to quote from materials in their possession: The British Library; The Royal Society of London; H.M. Customs and Excise; Dorset County Record Office; The Royal Greenwich Observatory.

Biographical

Very little is known about William Hobbs. He published nothing in his lifetime, and little independent record of him seems to have survived beyond his own unpublished natural philosophical writings. To some extent his life has to be reconstructed from these writings, though they contain little direct autobiographical information. The surname Hobbs was common, and late seventeenth and early eighteenth century local records in Wessex and the south-western counties reveal snippets of information about many William Hobbs's, it being impossible, at this stage, positively or negatively to identify some of such references with the author of *The earth generated and anatomized*.

It is not clear when or where Hobbs was born. In a letter sent to the Royal Society in 1709[1] he claimed to have been making natural observations for more than 30 years, which must surely put the year of his birth back before 1670, possibly considerably before then. There is no unequivocal clue as to his upbringing, education, or vocational training. It seems very unlikely that he was university educated, partly because he himself draws attention to his lack of skill in languages,[2] and partly for other reasons which will become apparent.

There were two excise officers with the name William Hobbs—they were father and son—working in the county of Dorset at the end of the seventeenth and the beginning of the eighteenth century.[3] It seems plausible to suggest that one of these—almost certainly the father—was the author of this treatise. An excise officer with an outdoor 'Ride', who would frequently be moved about the country by the Board of Excise, would have had excellent opportunity for gaining a knowledge of the structure of the countryside. Furthermore, a Dorset excise-man would have been expected to possess that kind of knowledge of the tides which the author proudly displays.

Both William Hobbs's, the excise men, were dismissed from the service—in January and April 1705 respectively—for dishonesty. One was subsequently reinstated, and posted to Devonshire.

My next—possibly appropriate—piece of independent biographical evidence dates from 1716, when the Weymouth and Melcombe Corporation Records note that a William Hobbs, 'Marriner', married one Rebecca Welstood, and soon found himself in trouble with his inlaws:[4]

William Hobbs of this Borough and Towne Marriner maketh Oath that about a weeke after Easter last past he this depon! intermarried with Rebecca Welstood of this Borough and Towne who was then possessed of a Messuage Burgage, or dwelling house in the [illegible] of Weymouth within this Borough and Towne called and known by the name of the Compass Alehouse and haveing so intermaried this depon! and his wife lived in and enjoyed the said house And this depon! saith that yesterday in the evening Edmund Welstood of this Borough and Towne Blacksmith Robert Saxton the elder of this Borough and Towne carrier John Williams of this Borough and Towne labourer and Anne Welstood of this Borough and Towne spinster ranne into the said House where this depon! then were in possession and after calling this depon! if he [illegible] severall names threatened to beate this depon! if he did not go out of the said house and this depon! saith that he went out accordingly and hath continued out of possession of the said house ever since and this depon! saise that he is afraid of going into the said house fo feare of being beat out of the same haveing been threatened to be so used

Jurat 10 of July 1716   William
                                                                               Hobbs

Of course, it is perfectly possible that the 'marriner' referred to is a third William Hobbs, who has no part of our story. If that is not so, however, it is more likely, I believe, that the Hobbs who fell foul of his new inlaws was Hobbs senior rather than junior, for Hobbs junior later seems to have had a wife named Hannah (see below). The outcome of the fracas of 1716 is unrecorded, but the bad feeling between Hobbs and the Welstood family evidently continued, for in 1722 we find in the Corporation Records:

3 April 1722 Ann Welstood, Singlewoman of this Town maketh Oath that On this present day Willm Hobbs, Seaman, inhabiting in this Town came to the House of this Deponent and Endeavored to Enter it, but she haveing Lock'd the Doore he broke her Windows and Unhung the Casement and Shook her through ye Window Sev.^{ll} times and swore he would kill her this Deponent, and then threw severall large Stones into the house at her and called her Whore & Bitch and other Scandalous names, wthout haveing any provocation given

Jurat coram majore
3 Ap^ll 1722   Ann Welstood

> Wm Hobbs tenetur in
> Thos Carter
> Mary Hobbs

And then

> 10 July 1722
> Mary Hobbs deposes she was assaulted by Anne Welstood
> [Mary Hobbs being, I believe, William Hobbs, *père*'s, daughter by a first marriage and
> thus William Hobbs, *fils*'s, sister.]

At about the same time there is a further record in the Weymouth Corporation Records of
a William Hobbs being employed as a school-teacher.[5] It seems plausible to suggest that this is
not the same querulous Hobbs who had been described as a mariner and seaman. In fact, we can
be pretty sure that this Hobbs is the William Hobbs who left a will in 1743, and who was the
brother of Mary Hobbs and the husband of Hannah Hobbs:

> This is the last Will and Testament of me William
> Hobbs late of Weymouth and Melcomb Regis in the
> County of Dorset Schoolmaster, and now of Osmington,
> made and published the fifth day of July in the year
> of our Lord one thousand and seaven hundred fforty and
> three And first I do make and ordain my Daughter
> Ann Hobbs sole Executrix of this my last Will and
> Testament, and my will is that she shall and do pay
> unto my Sister Mary the wife of Nicholas ffar of
> Chatham two Guineas and a half, In consideration of
> her kindness to me when I was last there, All the
> rest residue and remainder of my Goods & Chattles
> Rights Creditts Sum and Sums of Money due and owing
> unto me on any Account howsoever any personal Estate
> whatsoever I do give devise and Bequeath unto my said
> daughter Ann Hobbs IN TRUST nevertheless and to the
> Intent and Purpose that she shall and do joyntly imploy
> the same with her Mother, my wife Hannah Hobbs, in the
> best and most frugal Manner they can for their joynt
> Support and Maintainance during all such Time she
> shall and do continue my widow and no longer. But
> in case my said daughter shall and do happen to
> decease before my said wife that then I will and
> hereby give and devise all my said Goods and Chattles
> and personal Estate, so bequested in manner aforesaid,
> onto my said Wife, Hannah Hobbs, In Witness whereof

I the said William Hobbs have hereunto set my hand
and Seal the day and year above written.

Signed sealed published and declared ⎫
by the said Testator William Hobbs   ⎪
as and for his last Will and Testament ⎪
in the presence of us who as witnesses ⎬   Wm Hobbs
hereunto have subscribed our Names in ⎪
the presence and at the request of the ⎪
said Testator   ⎭
    John Fooks
    Joan Watts
    Edward Fooks

Hannah Hobbs Executrix in trust was sworn the 19th day of
August before me, J. Preston—as also that the effects of
the deceased did not amount to Twenty Pounds—[6]

The late date of the death of this Hobbs—1743—probably indicates that it was the son and not the father. Further evidence is provided by the signature attached to the will, which is markedly different from the signatures on the letters to the Royal Society in 1709. Hence, tempting as it is to believe that the Hobbs who was the author of *The earth generated* would have made a suitable schoolmaster for Weymouth, and with his obvious familiarity with navigation, geometry and mathematics, it seems far more probable that it was his son who became Weymouth's schoolmaster, whereas the author perhaps made his living as a mariner, and then (presumably) as an inn-keeper. It would not be unusual, one imagines, for an ex-excise officer to marry an inn-keeper's widow.

So far as the evidence allows us to judge, Hobbs seems to have lived a life of almost continuous isolation from the learned and philosophical communities of his day. Apart from the brief contact with the Royal Society and a fleeting correspondence with Flamsteed at the Board of Longitude, he appears to have kept at a distance from London societies and scientists. I have found no evidence that he had any contact with any of the other virtuosi, country gentlemen, parsons, etc., who lived in Wessex and the south western counties and who pursued similar interests—men such as William Cole (d. 1701), John Strachey (1671–1743), John Aubrey (1626–1697), Joseph Glanvill (1636–1680) or John Beaumont (d. 1731). Similarly, beyond the treatise, and the shorter papers to the Royal Society on much the same subjects, Hobbs does not seem to have left record of pursuing any other intellectual interests.

Our first reference to Hobbs's scientific interests comes from two surviving letters of 1709 to the Royal Society, in which he gives his address as 'near the Bear Inn, Weymouth'. It would seem as if Hobbs submitted in person to the premises of the Society at Christmas 1708 a lengthy paper (now preserved by the Royal Society) with sections on the theory of motion, the tides, trade winds, etc., together with an accompanying letter, which he had apparently hoped to be allowed to read to a meeting of the Society. The letter[7] runs

Read 11 May 1709

Hono:^ble S^irs

Before we proceed to my Intentions its my indispensible duty to make an humble Acknowledgemt as well of my unworthyness to appear before, as of my unfittness to spake in the presence of soe Honble soe Wise and soe Learned a Society; But that my Presumption may not be esteemed inexcusable, nor what I have to propose seem altogether needless, I begg leave to give your Hono:^rs the Reasons that induced me hereunto. In order to which be pleased to know that for above 30 years I frequented my thought as well in Mathematicall knowledge As in Mechanicall Curiosities; But observing of Late such Vast improvements made in the former

[here ends the first page, at foot of which is written 'ordered to be registered']

I became less thoughtful thereof, and diverted my Contemplations to the search of natural Causes, whereupon those Books that came to my hands relating thereunto, I read with great delight, and as greatly Admired the Learning and Wisdome of the Authors; But having compared their Doctrine and Hypotheses with each other, And those again with the things that Appeared, I found Contradictions therein, That I could not obtain the least Satisfaction in what I sought after; And therefore I justly concluded, That the Truths of Nature were not as yet fully discovered.

And this I the rather supposed, because the verry Persons who writt such Admirable things in Mathematicall Sciences, And therein Exactly agreed with each other, were the same that soe much differed in their Apprehensions concerning the things of Nature; From which Considerations I presumed to lay aside as well what the other had proposed, and to try if I could hit on any other foundation that might better Answer the Ends designed: In order whereunto I took a compendious View of the process of Nature in generall, Wherein I found (as Yo^r Hono:^r well knows) That y^t First Created Bring of which the Univ^rs was made was Matter only.

2^ly That this matter was by the Maker thereof Imprincipled with a Plastick power or faculty, to Forme it selfe into Naturall Bodys of Diverse Kindes and Species.

3^ly That this Formation could not be Effected without a generall Motion in all its parts. For if Matter, or any part thereof, had been always devoid of Motion, it had perpetually remained such, And Consequently, noe naturall Body could have been formed Wherefore all naturall Bodys are Educed out of Matter by Motion.

4^ly That altho' this Motion or plastic Faculty is originally promiscuously mingled with, and imprincipled in Matter as aforesaid; Yet the Matter of all such Bodyes (how promiscuously soever originally mingled) is nevertheless in its Forming, Tripartly Divided, or Discriminated; Namely into Active, Passive and Fixed parts.

5^ly That the Active part (when formed) gives Motion to the passive part thereof. And the Fixed Contains, Setts bounds unto, and regulates the s^d Motion, and also constitutes or modifies the Body soe produced.

6^ly That this Active part being thus Divided or Discriminated, is allways internally disposed of, or inclosed within the Body soe produced as afores^d, And not on its Superficies, or at a greater distance from it.

From these and other like self-evident Truths I grounded, or discovered my Hypothesis, By which I soon found That the Motion of the Fluid or passive part of the Earth, was not

Caused by the Motion, or pretended Influence of its Satellite, but in such manner as in the following Lines is Assigned. Nay it soe well Agreed with the Phenomena's of Nature, That if I saw but a Stone in the Street, (having any naturall Vain or Impression in it,) I could discover when, and how, it was produced or Impressed therein. Being by the same Doctrine also taught, That as there is a mutuall Harmony in the Powers, Proportions, & Affections, of Lines and Numbers in Mathematicall Knowledge, soe there is likewise as regular and mutuall an Agreement in the Powers Proportions and Affections, in the things of Nature; Which when fully discovered, and a true and Genuine Foundation Laid, I doubt not, nay am well assured, That the Doctrines, Corollaries and Consequences, Drawne from them will (from the Same naturall products and Phenomena's) be as evidently proved and demonstrated, and Errors as readily Detected, as by any of the Problems and Propositions in the Mathematicall Sciences whatsoever. And Consequently there will be as great concord and agreement amongst those that shall Treat thereof, as is now amongst those that treat of the former: For if such Regularity and perfection can be found in Art, which had its Originall from Humane Inventions and Institutions. How much more in Nature, whose Ordinances were Established, and whose footsteps were Imprinted, Long before those Learned Inventions, or humane Arts had their Being; And whose Laws are immutably the same, as having the Externall Existence for their Author: But if it has hitherto been otherwise, it's for want of a right Foundation And not from any deficiency in the Demonstrations that may be drawn from the Appearances or Evidences aforesaid.

Now by what hath been already delivered I have plainly intimated That the various opinions of those that treated of naturall causes, was what induced me to lay my Conceptions before yor Hono^{rs} in hopes at least that Something therein may tend to the Discovery of Such a true and Genuine Foundation as may produce such an Harmony amongst those that shall build thereon as is before proposed.

I would add a Word or two in particular concerning the following Lines, In which I have briefly treated of, And Assigned the Causes of the Various Motions in the Elements of Air & Water, i.e. of the Trade Winds, Monsoons And of the Flux, and Reflux of the Ocean.

In accounting for which (as has been already intimated) I have made some Angles with what severall of the Learned and Ingenuous [sic] of this Age have proposed. And altho' I am well Assured of the Truth and Certainty of what is therein Contained, Yet am nevertheless Soe unwilling to Give the least offence to those Worthy persons that have soe ingenuously [sic] treated of the same things Or to offer any thing to this Honble Society that may be judged needless of which they are fully Assured, the reall Causes are already Discovered: That I had rather Conceale what I have prepared and presume noe farther, than by a superfluous proceedure to be Guilty of offending in manner aforesaid. But on the other hand if your Hono^{rs} Thinke that any of the Phenomena's in the Motions of either of the said Elements, or in the Fixed part of the Earth, are not as yet sufficiently accounted for, and shall please to Admitt of my Endeavours, I shall freely Offer to your Consideration what I have prepared. And tho' I am not accomplished with such Learning and capacity as may be necessary for its Verball Justification, Yet inasmuch as they are not Founded on Speculative Notions, nor Drawn from remote powers and Influences:

But from the Visible and Tangable Appearances of Nature and Evidenced and Illustrated in such Manner as this Royall Society Expects and Requires, I mean by Experimental Examples and Observations; I Assure myselfe they'll not only obteyn yo.r Hono.rs Approbation, But will be improved to a far Greater perfection by the Addition of more profound Discoveries, Than by mean Capacity can pretend unto. Which that it may is the hearty Desire of yo.r Hono.rs most humble Serv.t

W.m Hobbs

Hobbs's letter is followed by his paper, 'An essay concerning Motion', which was read to the Society at its meeting of 11th May 1709. The Royal Society's Journal Book runs[8]

Papers of M.r Hobbs from Weymouth were Delivered by the President they related to the Structure of Mountains, Motions of Tides &c M.r Hunt was Ordered to take Care of the Modells and Stones and the Papers were Ordered to be Copied into the Books of the Society and he was Ordered to be thanked.

Presumably about this time Hobbs was sent the letter of thanks (of which no trace seems to remain) to which Hobbs refers in a later letter.

On 25th May William Derham (1657–1735) gave the Society an account of the paper Hobbs had sent. This paper survives in original and in copy in the Royal Society, but it is hardly worth transcribing in full here since it duplicates almost totally the ideas set forth in *The earth generated and anatomized* (though see later, p. 19 for an analysis of these ideas). Rather it is convenient to give Derham's account of it, since it is a reasonably accurate resumé, and embodies the judgment—prejudiced obviously, but not unintelligently—of the Society. Derham's report is headed 'An Account of the Papers Models and Figures which the Society Received from Mr Hobbs of Weymouth in Dorsetshire by the Reverend Mr Derham F.R.S.'[9]

In pursuance of the Orders of this Most Famous Society I have with as much Care in my Opinion as they Deserve read over Mr Hobbs's Papers, in which I find him to have a Clear Head, and to be a Person of great Diligence and pretty good Judgment considering his Want of Learning, But his Philosophy is much inferior to his Observations of Matters of Fact, And therefore altho' I don't think him to be Numbered among the Eminent Philosophers and may have a meaner Opinion of his εὑρηκας than what he seems to have himself, yet I imagine he may be of good use to the Society, if they should have any Queries to be Answered or Notices to be taken in the Place where he lives. Particularly I think his Observations of the Tydes may well Deserve the Cognisance of the Society, he having Observed them strictly for two years, and no doubt will if desired proceed with Delight in farther Observing them if it be thought necessary.

As to the particulars of his Papers a short account may Suffice Except where his Models and Figures may call for Some Enlargement.

The Causes he assigns for Rain have nothing new or Remarkable except what is D.r Woodward's Centrall heat.

The same for Winds. The generall Trade Winds he ascribes to the Diurnal motion of the Earth and heat of the Sun. He is of opinion that the Whirling about of the Terraqueous

Globe causes a Wake or Ditch in the Air between the Tropicks like that behind a Ship under Sail in the Waters, That the Hills and Highlands for the Divagations of the Trade-Winds near the African and the other Shores bordering on the Ocean.

For the Illustration of this he recommends to draw a piece of Wood thro' the Water, behind which you will not only see the Wake but you may observe also that just behind it the Water runs almost across it. So he says the Highlands next the Ocean whirling thro' the Air, the Wake in the Air next them is not (as 'tis at greater distance) Easterly & Westerly but toward the Northern & Southern points. And for the Monsoons he accounteth for them in this manner, He sayeth if instead of One you take two pieces of Wood, and brace them to stand at a distance from each other with a Wire bent so as to stand above them out of the Water, that there will be a Wake behind the hindermost but none between the two pieces of Wood, after the same Manner he imagineth it is where the Winds blow one half of the Year One Way, the other half the Contrary way (viz.̣) that the Wake in the Air between two Lands lying near One Another is taken off, as between Madagascar and Africa, between the Malacca, Sumatra, &c, which he thinks near enough even to Africa itself to cause this Quietness or Calm of the Wake of the Air. The force of the Trade Winds being thus taken off, he thinks then that the heat of the Sun takes place, and as he is either towards the N or S. Tropick, so he saith he drives the Vapours before him toward that Pole he is nearest unto, and then the Trade Wind Vapours come tumbling in thereupon, and make either an Easterly or Westerly Monsoon.

His Experiment and Notion I confess seemed to me Ingenious, and altho' I scarce think that the Rotation of the Earth has much to do in the Trade Winds, yet if his Notion be refined and considered upon it may be so far Serviceable as to give some good hints to Ingenious Persons about the Difficult Phænomena of the Monsoons, for which reason I have Enlarged on that head.

The last thing he Treats of, and thinks he has fully resolved, are the Tides, which he thinks can never be owing to the Influence of the Moon, or any Cælestiall Body, he knows nothing of the business of the gravitation of the planets to one another, and makes all Influence to be no other than the heat of Such Heavenly Body, and that the heat of the Moon or any Heavenly Body but the Sun cannot reach the Earth he endeavoureth to prove from the proportions & Distances which he gives of the Earth & Heavenly Bodies in his 3^d Scheme, In which Scheme he makes the Atmosphere and the heat of the Sun to be the Same, or at least that his Atmosphere Extends as far as his heat, as is represented in that Scheme by the Yellow Circular Shade about the Sun, and that the Moon hath no Concern in Agitating the Waters he thinks he has Demonstrated from this.

1st That no Influences can be conveyed from the Moon to the Earth but by some Medium or Atmosphere, and consequently the Atmosphere of those two Globes not Approaching each other as also they being both Earthy and Cold Bodies no Influence can be conveyed from one to the other.

2^d That the Motion of the Tides doth not Correspond with the Culminations of the Moon, which he supposes to be every Revolution only about 48′ different according to which Supposition he has drawn up his Table on the backside of Fig: 1. Whereas he findeth the daily Variation of Tydes to be at greatest Spring Tides but about 29′ & at the lowest Nepes 96′ and he Sayth he hath for two Years Observed the Variations of the Tydes

in the Sea near the place where he lives, and he had very Ingeniously and handsomely Contrived, and put them into Fig 1st and 2^d

Having thus as he thinks discarded the Moon from having to do in the Tydes he assignes what he takes to be the true Cause, and that is a kind of Respiration within the Earth owing to some Rarefactions therein, which he proves have been, and thinks therefore always are & will be in it as being as necessary for its Conservation as Animal Life and Motion is to the Conservation of Animall Bodies.

That Such Rarefactions have been in the Earth he proves thus, with Dr Woodward he supposes the Terraqueous Globe to have been once taken to pieces, as the fossile Shells &c import, but he denies this to have been at the Deluge, when this Mish:Mash had subsided into Various Strata according to the Rules of gravity and become hard, he thinks the Hills were then raised or pushed out by means of some Internall Rarefactions, These Strata & the manner of their being pushed out he hath handsomely (if I mistake not) represented in his Models (N^o) 1, 2, & 3, which I took but a Transcient View of at the Society, Having thus proved an Elevation in the Terraqueous Globe by means of Rarefaction he takes it for granted that it as necessarily Continues as breathing doth in an Animal, and that this happens at certain times near the Nepe Tides (like a sort of Respiration) and that one Agitation at Nepe Tides is Sufficient for all the following Agitations of the Waters which are only so many Returns of that Elevation of the Ocean.

To illustrate the matter he Instanseth in a long Wooden Tube or Trough with Water put therein which being lifted up at One End will Cast the Water to the Other, and being let down again the Water will return and Ebb and Flow backward & forward at first more & by Degrees less, and the proportion it doth thus in the Tides he hath represented at the Bottom of Fig: 2^d & in a Small paper by itself in a straight line.

Another Illustration he gives us is by a large piece of Level ground of half an Acre or an Acre with a large Map of the Sea and Land drawn thereon the Seas to be represented hollow lower ground the Dry Land by rising up higher. An hollow deep pit being sunk in the Middle & covered with Leather & Water poured on the Leather if the Leather be lifted up in the Middle, 'twill Cast the Water out among the Seas and against the Dry Land and make such like Variations he thinks in the Ebbings & Flowing in this Map of the World as befalls in the Terraqueous Globe itself.

And lastly to further prove this Respiration of the Terraqueous Globe or to put it in his own Words; 'That there may be a Musculous part formed under the Ocean, where it may operate by its own principle or be agitated by some Internall Cause, or Rarefaction', *As his Words are p. 35*—I say to make good this he Instanceth in the Dilation and Contraction of the heart of an Eel for some time after 'tis cut out & but especially in the Elevation and Subsiding Motion observable in the Bell Fish which he hath sent a pattern of to the Society.

As to his Notions about the Motion of the Heart of Animals, & many other things, which I have passed by, I thought them not worth the Cognisance of the Society neither indeed should I have said so much of most of these as I have done but only to acquaint the Society with the Design & Meaning of his Models and Figures which are pretty enough to recommend them to the Desires of the Curious to be satisfied about them.

But reviewing his Papers there is One of his Figures I find I had like to have passed over, and that in his handsome Draught of the Isle of Portland, and his Remarks on it's Tydes. 1.st He Observes when the Tydes in the West Bay set Eastward that in the long Lake or Lagune (Setting in at the Narrow Passage B) they set to the Westward, and that 'tis full Sea at C, three hours later than at D tho only a Furlong distant. 2.^{d} that in the Bay A, the Water ebbs & flows many times an hour or two before the proper Flood, which preliminary Tide they call the Gourder, this he applies to what is reported of the Tydes near Negropont. 3.^{d} That at B, the Water oftentimes runs in briskly thro' the Inlet altho' it be Ebbing Water and Sunk a Foot or more in other places, all which things I think are easily Accounted for by an intent Inspection and Consideration of his Scheme.

Derham's account of Hobbs's papers drew some discussion. 'Doctor Woodward said that his Observations on the Strata of the Earth were Conformable to his Theory, and that he had given Mr Flamsteed his Observations on the Tides which differed from these given to the Society.

Mr Derham was desired to compare Doctor Woodward's or Mr Flamsteed's Observations on the Tydes Delivered from Mr Hobbs with those Delivered by him to the Society.'[10]

This then led to a further exchange at the meeting of 13th July when 'Dr Derham Delivered the Papers of Mr Hobbs Concerning the Tides and Earth, which were Ordered to be putt into the Repository after being Copied in the Books and returned from Mr Halley who was desired to peruse them and give his thoughts of them to the Society.'[11]

Halley apparently never did so, and no further mention is made of Hobbs in the records of the Society.

Hobbs meanwhile was clearly becoming anxious as to the verdict of the Society on his works, having had but a bare acknowledgement of receipt. Hence he wrote a few days later.[12]

near y^{e} Bear Inn in Weymouth
July 16: 1709

S^{ir}

I rec.^{d} yor obliging Letter wherein you were pleased to give me An Account That the Papers &c; which I left with the Hon$^{ble:}$ Sir Isaac Newton at Christmas last were by the Honble Society ordered to be Registered: for which I retourn my humble thanks. I should have been glad of a Line whether any part was approved or if not wherein Rejected. I will assure your Honours that 'twas neither Profit, applause nor emulation that enganged me thereunto But the irreconcilableness of what I saw in naturall things, to what I Read in those that have written thereof. And though my Pen be not sufficient to establish what I therein attempted yet perhaps Posterity may be by some Learned Author convinced that the Rocks were not Dissolved and the Shells immassed therein at the Deluge 2:^{ly} That the Moon is not the Cause of the Flowing of the ocean, and 3:^{ly} That the Diurnall Rotation of the Earth is the principall Cause of the Trade-Windes, Monsoons, &c Pardon me S^{ir} for this Freedom tho' you should be one of those that approves of what has bee [sic] by some Hono.^{ble} and Learned Authors written to maintain the Contrary. S^{ir} I have seen soe many Thousand Tonns of Stone soe plentifully immassed with Shells, And made such observations of the generall Scituation of the Stratus or (rather) Bedds of Earth And kept

soe long Account of the Flowing of the Ocean where noe river hindered That the above particulers are as plainly deducable from these observations at that the Sun is the fountain of Light &c: (or otherwise I should not have soe presumed.) And this I purposed to have evidenced by Experim^ts before yor Hono^r: but was unhappyly prevented by your not meeting in the Hollidays.

I have herewith sent you 2 or 3 minnerall Impressions of which I have not seen any till lately. I have others in store and if Acceptable shall freely present them—Please to Hono^r me with a Line of the Receipt and if not too much trouble one line to the first above mentioned in which you will oblige.

> Hono^rd S^ir Yo^r most humble
> Servt
> Wm Hobbs

I took the opportunity of a neighbour to carry it gratis

There Hobbs's contact with the Royal Society seems to end. There is no evidence that when his ideas were read out to the Society, they created any interest. Nor is there much sign that Hobbs thereafter tried to make further contact with the London intelligentsia during the rest of his life.

The next record of Hobbs is his treatise itself, *The earth generated and anatomized*. Of this, the preface 'To the Reader' is dated 1715 (though this date has almost been scratched out). It is very difficult to date the original composition of the treatise. Its Postscript leads one to believe that far the larger part of the treatise was conceived and in some form written several years before the Postscript. It is possible that Hobbs's failure to interest the Royal Society in his papers led him to decide to reorganize his materials into a lengthier, more coherent, independent, treatise. There is no clear indication whether Hobbs had any intention or ambition to publish the treatise, though it is written as if addressed to an audience (rather than being a series of self-addressed notes).

The treatise was composed in Weymouth. Hobbs had clearly been domiciled in Weymouth in the early years of the eighteenth century, since the treatise contains a register of the tides at Weymouth kept by him then over a period of three years, which may well indicate an ambition to compose a theory of the tides and of the Earth over a decade before he contacted the Royal Society.

William Hobbs in his local milieu

William Hobbs lived at a time before provincial England generated on any regular basis scientific communities and scientific societies. The period saw of course many provincial naturalists, but they were essentially *sui generis*.[13] None of the other naturalists operating at this time in the south western counties of England—men like John Strachey at Chew Magna, and William Cole at Bristol—lived sufficiently near for Hobbs to be part of their circle. Dorset in Hobbs's day does not seem to have supported any kinds of gatherings amongst men with scientific interest.[14]

Weymouth itself, being at the very southern tip of Dorset, was particularly isolated. It was over a hundred miles from Oxford and further from London. Having suffered under siege

during the Civil War, Weymouth was a small, and probably decaying, sea-port at the turn of the eighteenth century. Its growth industry was smuggling, mainly from France via the Channel Isles. Weymouth did not even possess an endowed grammar school. It is hardly mentioned in the various books of travels which were produced early in the century.[15]

Weymouth's rise to some degree of importance as a town dates from the second half of the century, when it became popular as a seaside resort as a result of George III having chosen to use the town for sea-bathing.

If Hobbs received little human stimulus, however, from his Dorset milieu, he, like many a later geologist, must surely have been deeply stimulated by the opportunities offered—indeed the scenes thrust upon him—by the local natural scenery. Dorset contains a fine succession of strata from the Chalk of the Cretaceous down to the Lias of the Jurassic, particularly well exposed by the extensive cliff scenery around Lyme Regis on the coast. The Portland stone quarries doubtless helped to convince Hobbs not merely of the depth to which fossils were embedded in solid rock, but also of the fact that the Portland stone was largely composed of organic remains. The hills and cliff scenery of the county convinced Hobbs of the tilting of the strata in upland areas. Chesil Beach must have focused Hobbs's attention on the interface of sea and land, on accumulation of materials and denudation as a crucial process of Earth history. Hobbs was the first of a long line of geologists whose work was deeply inspired by the relief and structures of the county, and—later—by the extraordinary fossils to be found in the vicinity of Lyme Regis. Indeed, Osmond Fisher, who wrote extensively on Dorset geology, was born at Osmington, which is where Hobbs—probably *fils*—died.[16]

The intellectual milieu of Hobbs's *The earth generated and anatomized*

I shall not attempt here a detailed account of the state of the science—or sciences—of the Earth in Hobbs's time. Detailed references and bibliography on particular areas of this subject will be found in the notes appended to the main text. Rather, I should like to bring into focus certain aspects of the ambience of contemporary science, particularly as it impinged upon a scientific amateur such as Hobbs.

William Hobbs lived in a world in which the modern divisions of Nature and scientific inquiry into physics, astronomy, chemistry, biology, geology, etc., did not exist as such. Hobbs thus characteristically had no generic or specific terms for his science. He saw no conceptual oddity in writing—nor would any of his contemporaries have had difficulty in reading and assimilating—a tract which related together extensive local observations of the Earth with a geocosmic philosophy; which attempted to understand the structure of the Earth's crust in terms of celestial philosophy, and in context of a total philosophy of Nature. He saw the terrestial world as analogous to the animate, and indeed believed that all Nature was governed by laws of animation, organization and generation. In this respect, Hobb's work does not represent an antiquated rag-bag of quaint juxtapositions. Such a conjoint range of interests are the common ones of the age, and are utterly typical, for example, of the work of Halley.

Indeed Hobbs had a good grasp of the main contemporary problems within the philosophy of the geocosm. Amongst the chief debates to which he addressed himself were:

(*a*) The debate over the theory of the Earth, which had been joined in Britain in the works, above all, of Hooke (1635–1703), Burnet (1635–1715), Woodward (1665–1728), Whiston (1667–1752), Warren, Croft, Arbuthnot (1667–1735) and many others. The chief issues were whether the Earth had been created, or was eternal; how it had come to assume its present shape—its landforms, the divisions of land and sea, mountains and valleys; the questions of its age and its likely future. Furthermore there were problems of a more specific and technical kind, such as the nature and history of rivers and fossils, the origin of soil, and the question of an inner heat for the Earth. On most of these issues, Hobbs felt obliged to argue for a particular case, and to dismiss others. On some of them, however, *e.g.*, the questions of the origin and nature of fossils, Hobbs clearly had no doubts, and did not feel any need to spell out the criteria for choosing between different interpretations.

(*b*) The debate over the theory of the tides. This was not yet fully solved by the end of the seventeenth century. Hobbs was not out of date in still puzzling over the issue. The major seventeenth century theories had been those of Galileo (1564–1642), who attributed tides to the Earth's rotation; René Descartes (1596–1650), who saw them as a product of pressure set up by vortex action; and various theories specifically of lunar attraction. Some of these latter (as for example Childrey's (1623–1670)) were somewhat astrological, or magnetic; some, as Wallis's, were mainly concerned to establish the empirical regularities between the moon's phases and the tides; and some, as Newton's, emphasized the agency of gravitation. This clearly involved much larger natural philosophical issues with which Hobbs grappled, such as the problem of ether, and void space; and of the differential communicability of light, heat, fire, and power *via* media through the solar system.

As with most of his contemporaries, and especially the contemporary 'common scientist', Hobbs held a melange of views. Some of these could be called 'modern', and were destined to have a future. Hobbs was, for example, a 'modern' in his extensive commitment to factual observation, experiment and to the quantification of the tides. He was strongly hostile to 'verbal' science, and to mere book learning; he fiercely repudiated astrology and other supposed arts (taking a 'continental' view, as it were, in rejecting Newtonian gravitational attraction as 'astrological').

On the other hand, in the great seventeenth century debate on the order and composition of Nature, Hobbs's philosophy was obsolescent. The Renaissance had seen a great recrudescence of philosophies of active Nature; of alchemical and chemical philosophies; of the belief that Nature was living and organized. Such ideas were being thoroughly displaced in the second half of the seventeenth century both by the more orthodox forms of Cartesian mechanical philosophy, which emphasized that Nature was only matter in motion; and by the more complex Newtonian philosophy, which retained a far greater role for God's sustaining activity in Nature and for Providence, within a general philosophy of atomism and the downgrading of matter itself before law-governed forces. In this climate, Hobbs held to a philosophy which saw all Nature as alive; a philosophy with some affinities to that of the Cambridge Platonists, but one which owed far more to an alchemical vision of the potentialities of Matter for self-organization and generation. Hobbs had no concern for the providential dimensions of natural law; he had no interest in the ontological and epistemological problems of primary and secondary qualities. In a world increasingly of Newtonian forces and Lockean powers, Hobbs's philosophy was left behind.

Resumé of the argument of Hobbs's *The earth generated and anatomized*

On the thirty-first page of his MS (p. 59 of this edition) Hobbs states with full clarity the scientific problem which it was the aim of his treatise to solve. 'Our principal design', he writes, is 'only to finde out ye manner how, and when, the Shells, and other Marine productions, came to be immasssed and mingled in the Rocks and Mountains' of the Earth. For, as Hobbs endeavours to show from the beginning of his work, there are certain natural phenomena, easily visible to the observer, whose very existence and situation are puzzling, and which need to be explained. These are the fact that dry land exists, above sea level; that this terrestrial matter is composed of rock which is predominantly found stratified; that these strata generally run parallel to each other and to the horizon, though, where the relief is undulating, the line of the strata is generally parallel to the dip of the land. Furthermore, Hobbs takes it to be a puzzle that this solid rock contains the remains of shell-creatures and of fish and other marine bodies embedded within it—and embedded deeply, not merely superficially; though scarcely any remains of land-dwelling creatures are found similarly immassed in the rock. How this came to be so Hobbs takes as the problem of his work. Furthermore, he states that, this being a work of *natural* philosophy, these problems must be solved *naturally*.

Hobbs first establishes that the present land contains marine materials. It is for him inconceivable and unnatural to suppose that the sea could somehow have become raised up all over the land, and thereby have deposited such materials (above all, fossils). Hence, it is evident that the present continents must once have formed the seabed. It is also clear that the silts, muds, clays and other materials which now constitute rocks must have been plastic, flexible, and soft while on the seabed; otherwise shells and other organic remains could not have been introduced into them—for it is evident that such organic remains were introduced while on the seabed. (Hobbs offers a variety of arguments, when tying up loose ends towards the end of his treatise, to demonstrate that there is no way in Nature whereby the Noachian Deluge could have infiltrated organic remains into solid rock.)

Hence, Hobbs notes, the problem is to discover the mechanism whereby seabed became land surface. This process, he avers, must have taken place in two separate stages, as the visible evidence proves. Firstly, dry land must have appeared while its solid materials were still flexible, thus explaining the generally horizontal lie of the strata. But hills and mountains can only have been raised above the general level of the land after the continents had hardened: for they show evidence of fracture, and fluid land materials would not have held their place. In short, strata are not the product of a Deluge, or of precipitation, but of *elevation*: elevation in two stages.

The natural philosophical viewpoint from which to understand this process of the elevation of strata is through a general philosophy of Generation. The generation of the Earth is merely one example of generation at work throughout Nature; and one instance of the animated character of all bodies. Hobbs sets out a number of Postulates which explain his philosophy of generation. All bodies are presupposed to move from an original condition of chaos and undifferentiation, into a state of being organized, differentiated, animated, and vital within a system. All animated bodies need, and have, Life, Heat and Motion. But they also need as well as these Active Parts, Fixed Parts and Passive Parts. In the case of the Earth these are, respectively, its Heart, its Land Masses, and its Waters.

Thus to solve the problem of the strata and of fossils we must understand the process of the generation of the Earth. The Earth began from a chaotic condition. Gradually differentiation of its original materials took place, as in the development of different parts of the substance of an egg and finally the chick. First to develop were an internal pulsating faculty, or heart; and watery fluids which covered the surface. The pulsation of the Earth's heart agitated the waters and gave rise to tides (tides thereby being empirical evidence of the existence of an internal heart to the Earth).

Hobbs here interposes two notes of warning. Firstly, though the Earth is thus generated, organized and animated, its organic parts cannot be read off mechanically by a macro-cosm/microcosm comparison with other animals. The Earth is a very special animated whole with its own unique organic functions. Secondly, Hobbs seeks at considerable length to demonstrate the falseness of all other theories of the origin and cause of the tides—the notion that the rotation of the Earth, or the influence of the Moon (whether magnetical, astrological or atmospheric-gravitational) produces the tides. All such theories of lunar influence are empirically false (*i.e.*, they do not square with the periods of the tides). They are also physically erroneous. Lastly, they are ontologically absurd, for the tides must be organically and necessarily related to the economy of the Earth, not an accident of a mere satellite. Indeed, the rotation of the Moon is itself a product of the rotation of the Earth which is a product of the action of the tides, which is in turn a manifestation of the animated pulse of the Earth.

The pulse of the tides was (and still is) from the Poles towards the Equator. Gradually, as the tides drove water across the pristine Earth, some areas of higher, and some of lower, seabed formed, with *stratum super stratum* of loose sedimentary material being heaped up in horizontal layers. Gradually, some of these strata appeared above the level of the lowest neap tides; and then eventually above the level of the highest spring tides. Thus land appeared.

This land gradually hardened as a result of the action of the internal heat of the Earth. As irregular continents of dry land formed, the tides begin to fall into disequilibrium, causing the Earth for the first time to begin to rotate on its axis. Thus time, as man knows it, began. In stagnant pools which abounded on the nearly flat surfaces of the land, terrestrial creatures spontaneously generated (thus explaining why no land creatures are to be found deeply embedded within the rocks). In due course, the internal heat of the Earth raised up some parts of the land into hills and mountains, causing disturbance of the strata. The humid, fertile condition of the Earth, favourable to spontaneous generation, now gave way to a brisk circulation of waters on the land surface, with the development of systems of rivers, fed by rain. The spontaneous generation of large creatures ended—now, only minute bugs generate in that manner. Thus, the Earth is completely generated, and in a state of perfect preservation.

Hence, concludes Hobbs, by adopting a philosophy of generation, we can understand the main puzzling terrestrial phenomena, *viz*:

(*a*)   that land surfaces are the product of tidal action;
(*b*)   that tides created the rotation of the Earth;
(*c*)   that hills and mountains were raised up by the internal heat of the earth;
(*d*)   that the internal heat of the earth, together with the fertility of the original chaotic materials, brought forth and continue to sustain life;

(*e*)   that only *marine* organic remains are found fossilized in rocks;

(*f*)   that most rocks are horizontally bedded;

(*g*)   whereas in the vicinity of mountains and hills the strata are not horizontal but rather broken and ruptured in line with the line of the hills.

We also understand through this philosophy of generation that the Earth is not eternal, but rather had a beginning, and hence a Creator who is to be praised and worshipped.

To this treatise Hobbs then adds a Postscript, proving from empirical materials that the impulse of the tides is from the two poles of the Earth, thus reinforcing his argument that the Earth has a heart, or pulsing faculty, situated at both these poles.

William Hobbs: other scientific writings

Apart from *The earth generated and anatomized*, the only extended scientific paper by Hobbs known to the present editor is 'An Essay concerning Motion'. This Hobbs appended to his first letter to the Royal Society, read 11th May 1709, and it survives in the original (Royal Society MS Ex. 1. 13) and in copy (Royal Society Letter Book Supplement, G–H, copy, pp. 357–406). It is the paper summarized and criticized by William Derham on 25th May 1709 (see earlier). Practically all the major arguments and themes developed in the 'Essay' are also present in *The earth generated*: a discussion of winds, of tides (and of the impossibility of the Moon being their cause), and a demonstration that the cause of the tides was an internal pulsation of the Earth. For this reason it will not be necessary to discuss the 'Essay' at length here. The 'Essay' contains a few examples and experiments which do not appear in *The earth generated*. Conversely, very little of the lengthy discussion of the philosophy of generation, within an alchemical context, which is such a prominent feature of *The earth generated*, finds a place in the 'Essay'.

The most important distinction, however, between the two works is their focus of organization. The 'Essay' centres on a discussion of the various types of motion found in Nature. *The earth generated* is expressly concerned with an explication of the various phenomena of the globe in terms of its own unique history. To what extent this shift in focus represents a real change in interests, or simply the reorganization of ideas onto a different plane of coherence, is not clear.

The 'Essay concerning Motion' begins with a threefold distinction of Motion, into artificial, natural and accidental. It establishes that all motion is produced 'either by Rarifaction and Condensation, Sublimation & Precipitation, Dilation and Contraction, or by Causes derived therefrom or depending thereupon'. Artificial and accidental motion are quickly dismissed as falling outside his present concerns. Natural motion is then taken and divided into (*a*) intrinsick (*b*) internal and (*c*) local. Intrinsick motion is taken to be the motion of the Chaos 'during the time of its Incubation' before matter was divided up into three parts, active, passive and fixed. Internal motion is the motion of the parts of an organized and living body. Local motion is the movement of bodies from place to place in the universe.

Hobbs then addresses himself to the problem of explaining various particular sorts of motions associated with the terraqueous globe. Rain is briefly dealt with first. It is shown to be a motion consequent upon the formation of clouds, which in turn depends upon the action of heat in sublimating the 'watery humidity' of the atmosphere. When clouds achieve sufficient 'magnitude and Ponderosity', they 'begin to precipitate towards the Earth'. Then Hobbs examines winds. He is interested in the question of why winds exist at all. He recognizes that heat (particularly the heat of the Sun) creates a motion in the atmosphere—with the sublimation of the watery content of the atmosphere, the formation of clouds, and their precipitation as rain. But such heat on its own would produce merely vertical movements of the air, up and down, not winds blowing across the globe.

In part, he concludes, winds are created by the inequalities of land and sea, which cause certain parts of the Earth's surface and atmosphere to be hotter and cooler, rarer and denser, than others, leading to movements tending to equalize heat and pressure. But chiefly winds are the product of the Earth's motion, both diurnal and annual. Hobbs assumes that the atmosphere drags and is perturbed as the Earth rotates on its axis. He believes that the Earth's rotation accounts nicely for the fact that the fiercest winds, such as Monsoons and Tornados, and the most constant, regular winds, the Trade Winds, are found towards the tropical, rather than the polar or temperate, regions of the globe.

For in the equatorial regions, the Earth is spinning with greatest velocity on its axis. Towards the poles the Earth's motion is far less; hence, winds are much weaker. Similarly, the direction of Trade Winds (towards the Equator, and westerly) is explained by the easterly rotation of the Earth. In his discussion of winds, Hobbs makes extensive use of Halley's views and of his map of the Trade Winds published in the 1705 edition of his *Miscellanea curiosa*.

Hobbs then proceeds to discuss tides. His argument covers essentially the same ground as in *The earth generated*, but is actually expressed with greater clarity and economy. He sets out his basic premise thus:

> In the first place, then, I shall lay down this Proposition; namely, That the Waters of the Great Ocean are by some internal Force or Rarification within the Cavity of the Earth, in some certain part thereof lifted up or exsusitated above its Equilibrium, or equal distance from its Center, & yᵗ it being so rais'd, does from thence naturally devolve towards the Shore, whereunto being come, the Waters are there made higher than where they were first lifted up; and therefore they as naturally return to the place from whence they first came, which thereby does again become the highest part of the Ocean; on which it again devolveth as before. And in this same manner with the first Elevation, it Ebbs and Flows for fifteen Days successively, at which time by receiving another Elevation, it Ebbs and Flows again in the same manner as aforesᵈ.

In order to substantiate this view, Hobbs undertakes to prove three further postulates:

> First, That there was, and
> Secondly, That there is, such an internal Force within the said Cavity or Bowels of the Earth.

His proof of this is largely from the existence of mountains. The only Power in Nature which could have created mountains is an elevating heat-force located somewhere beneath the Earth's crust. Such a force, which once existed, presumably still exists to convey impulse, or wave motion, to the tides; for, argues Hobbs, by the analogy of Nature those powers which assist in the *formation* of organic beings continue to ensure their *sustenation*.

Despite popular opinion, Hobbs avers, tides cannot be caused by lunar, stellar, or solar attraction (any more, indeed, than mountains could have been raised by such supposed forces). Hobbs maintains the point in two ways. Firstly, he asserts that force can only be communicated by contact action in a sufficient medium. But no adequate medium exists between the Earth and the Moon. Secondly, he asserts (referring to his table of the tides kept at Weymouth) that whereas the theory of the lunar causation of the tides requires that the retardation of the time of the tides should keep pace arithmetically with the retardation of the Moon, in fact the retardation of the times of tides obeys a law of its own, which is not at all synchronized with lunar times.

Hobb's third postulate is that

the Ebbing and Flowing of the Ocean is exactly conformed or assimilated to Water put in motion by being so lifted up, & agitated as aforesaid.

He seeks to demonstrate this by setting up two experiments. One is the experiment with the pit in an orchard covered by a leather tarpaulin, described above in the account by Derham (see above, p. 14). The other is the experiment with the double-ended chute (designed to show how the height of waters originally raised at one end of the chute, and then allowed to pass from side to side, diminishes very gradually), which appears in *The earth generated*.

Hobbs lastly attempts to explain anomalous tide phenomena, such as the fact, demonstrated by Halley, that tides flowed only once every 25 hours in the gulf of Tonquin. He accounts for these by arguing that in enclosed areas (such as the lagoon behind Portland created by Chesil Bank) the frequency of tides is totally distorted by the accidents of the coastline.

The other scientific contact which Hobbs is known to have had is with John Woodward and John Flamsteed (1646–1719) concerning his register of tides kept at Weymouth. It is not clear how Hobbs made contact with either, nor when he did. There are two pieces of surviving evidence. One is a three page note in Hobbs's hand, headed 'A True Register of ye Time when Full Sea at Weymouth an^o Do^{nus} 1700 by Wm Hobbs communicated by Dr Woodward' which is now in the Flamsteed Papers at the Royal Greenwich Observatory (vol. 41, f. 129^v–130^r). The first two pages of this record Hobbs's observations of the times of the tides from April 1700 to October 1700. With a few exceptions—which I assume are slips of the pen—the data correspond to those in *The earth generated and anatomized*. The third sheet analyses the retardation of the times of the tides, and concludes 'By these diurnal Observations it appeareth that the mean Dayly difference of ye Coming of ye Tydes 2 or 3 days after ye New and Full Moon is but 27 or 28 Min. But 2 or 3 days after ye Full and last Quarter is [sic] 90 or 91 Min Difference W^m. H'.

The other is also in the Flamsteed Papers (Royal Greenwich Observatory, vol. 69, fol. 259^v–260^r). It is headed 'Mr Hobbs of Weymouth's Tables of the Tides observed by him there'. It is in the hand of Flamsteed's assistant, Thomas Faber (*alias* Smith). It consists of three geometrical diagrams graphically representing Hobbs's claim that his register showed that the time of high tide retarded by a logarithmic rather than an arithmetical law. The diagrams are headed:

'This scheme drawn from above two years observations shews how the flowing of the Ocean dos dayly increase and decrease from one Spring tide to another.'

'This scheme shews how the Tides should Flow if Governed by the Moon, which is contrary to Observation, as by the next Scheme appeareth', 'This Scheme drawn from above two years Observation sheweth that the dayly difference in the Flowing of the Tides is in a Logarithmicall, & not in an equall progressive Difference, as hath been in the past accounted.'

The only surviving indication of relations between Hobbs, Woodward and Flamsteed is a note in the Journal Book of the Royal Society which reads: 'Doctor Woodward said that his Observations on the Strata of the Earth were Conformable to his Theory, and that he had given Mr Flamsteed his Observations on the Tides which differed from these given to the Society' (vol. X, 1702–14, May 25, 1709).

William Hobbs: an assessment

(a) Hobbs's general vision of nature

It is difficult—and perhaps unnecessary—to try and pin down the exact sources or tendencies of Hobbs's fundamental vision of Nature. Its distinguishing feature is to see Nature as organic through and through. Hobbs places special emphasis upon Nature's vitality, its 'Vivifying, or self forming Quality', its 'plastick Qualification'. 'Nature is never idle', he writes. The power to generate is one of Nature's 'essential properties'. This general insistence on the essential quality of life possessed by Nature distances Hobbs utterly from the tendency of contemporary mechanical philosophy in England. It is also apparent in his stance over particular issues, such as his defence of Nature's capacity for spontaneous generation, and his hostility to the preformation theories which fitted in so well with the mechanical philosophy.

Hobbs's natural philosophy allowed no sharp distinctions between animate and inanimate. Yet his view of the activity and agency of Nature was hardly mystical, Hermetic, or built on the basis of a magical, or primarily symbolic philosophy—despite occasional mention of the mysteries and the symbolical characters of Nature. Hobbs was deeply hostile to 'vulgar' manifestations of mystical and magical views, such as astrology, or belief in fairy rings, just as he was hostile to introducing supernatural explanations into natural philosophy.

Rather, Hobbs was committed to what is essentially a chemical, or alchemical—the distinction is not material—view of the cosmos. His main concern was to see Nature in terms of such substances as Salt, Sulphur and Mercury, fermenting liquors, and menstrua. He was interested in states such as Heat, Motion, Fire. He perceived throughout Nature processes such as fermentation, germination, generation, incubation, concoction, separation, putrefaction, calcination, digestion, incorporation, vivification, and solidification.

In these respects, Hobbs's theory of the Earth addresses itself to quite different problems from the majority of the theories of the Earth of the late seventeenth and early eighteenth centuries. Hobbs for example did not at all share the common fear that a philosophy of Nature seen as active was the thin end of a wedge of full-scale hylarchism and hylozoism, which would lead to eternalism and atheism. Hobbs did not share the Newtonians' overriding desire to demonstrate the existence both of universal Law and Providence throughout Nature.

In many respects, Hobbs's theory was simply old-fashioned. In many others, it was desperately ignorant, wrong, ill-informed, naïve, and muddled in its views. For example, his arguments that the rotation of the Earth is the cause of the orbit of the Moon seem directly to contradict his claim—which he advances while when disproving lunar theories of the tides—that power and motion cannot effectively be communicated between the two bodies for lack of an adequate medium. It is not clear to what extent he was ignorant of many of the developments of contemporary astro-physics, or to what extent he merely ignored them. Thus, he did not explicitly discuss the Newtonian theory of the nature of the motion of the tides.

But we must not be hastily dismissive of the provincial's natural philosophy. For in many ways it was, in fact, highly appropriate as a tool for understanding the nature of the Earth, viewed as a system, as a complex economy, in which each aspect had its functions and structures, its part to play in the complex maintenance of the whole. Hobbs's was a philosophy excellently fitted for coming to grips with the phenomena of gradual change on the Earth's surface; highly appropriate for understanding the actions of the tides, the importance of central heat, of elevation, of denudation, of stratification—many of which phenomena Cartesian mechanical philosophy, or a Newtonian emphasis on particles and gravitation, were ill-suited to explain.

Indeed, it is important to point out that several of the theories which were to shape the development of geology in the eighteenth century shared many of Hobbs's own concerns and patterns of explanation. French writers such as de Maillet (1656–1738), le Cat, Boullanger (1722–1759) and Buffon (1707–1788) all, like Hobbs, stressed the importance of the tides as an agent in the formation of land masses. Like him they thereby naturalistically explained fossils and the elevation of land. As did G. H. Toulmin, in his *Antiquity and duration of the world* (1780) and Lamarck (1744–1829) in his *Hydrogéologie* (1802), Hobbs insisted on the synergistic unity in Nature of the organic and inorganic, with the organic having ontological priority. Like Buffon and Lamarck, Hobbs supported belief in spontaneous generation as a way of showing how the forms of life at any time were dependent upon the conditions obtaining on the earth.

It is easy to scoff at Hobbs's insistence that the Earth was an animal, and to dismiss his thinking for being crudely animistic. But it is important to remember that Hobbs himself reminds the reader that he does not thereby mean (unlike, say, Thomas Robinson) that the earth actually has ears, eyes, a mouth, hair, etc., but rather that the Earth is an organized system, in which functions correspond to structures, and which is essentially self-maintaining. When James Hutton (1726–1797) wrote in 1788 that the Earth was more than a machine—it was an organism as well—he was expressing, though in a rather more metaphorical form, essentially the same idea as Hobbs's.

(b) Hobbs, religion, and the aims of his treatise

Hobbs more than once affirmed that he was writing a tract of natural philosophy, to solve specific natural problems. This fact has two important correlates.

Firstly, Hobbs was throughout concerned with what was rationally and naturally possible in Nature. He termed himself on his title page 'a Lover of Truths drawn from Nature and Reason'. He did not of course absolutely exclude the possibility of miraculous, divine intervention. But he announced his hostility to invoking divine miracle (such as the Deluge) as a blanket *explicandum* of events which could be explained by natural causes. By implication and association, Hobbs linked the Deluge with the credence given by the 'vulgar' to such phenomena as fairies, and with Popish superstition. When Hobbs cites the Bible, it is to conform that God works by gradual, natural causation (*e.g.*, through the process of generation).

Secondly, Hobbs's treatise has no pretensions to be a total, man-centred cosmogony and cosmology, justifying the ways of God to man, as were, for example, the treatises of Burnet, Whiston, Ray (1627–1705), Woodward and other contemporaries. Throughout the work, Man is never mentioned in a context of design and teleology. Hobbs barely touches upon God's purposes in Creation. Moses, he asserts, did not seek in the Book of Genesis to offer a comprehensive natural philosophy of Creation. The aim of the Bible in respect of the natural world was merely to demonstrate that the Earth was not eternal; that it was God's Creation.

Nevertheless, Hobbs's world-view is Christian, and he seeks to show that the first chapters of Genesis (and also other Books of the Bible, such as Psalms and Job) do contain, when properly interpreted, natural truths. Yet his theory is in no sense merely an *explication de texte*. Far from it. For Hobbs declines to take a literal view of the Genesis cosmopoeia. He is insistent that the various processes of creation which are described in the first chapter of Genesis did not occur in six days, as the term 'day' is at present understood. They were rather gradual processes and must have taken a great deal longer. Furthermore, Hobbs offers a daringly unconventional interpretation of the first chapter of Genesis in which the six days of Creation do not even stand for six successive periods, but rather six synchronic aspects of Creation, all occurring essentially simultaneously, or at least being not significantly differentiated by succession. The difficulty of interpreting the six days as successive stages is that they show the differentiation of land and sea occurring *before* the creation of life. This seemed to exclude the most direct and simplest explanation of the entombment of fossils (taken as organic remains) in the strata, namely that the seas had been teeming with fish and shell creatures before the continents had come into existence. By asserting that the 'days' of creation were merely metaphors for different aspects of the creation process, Hobbs could employ this easy and natural solution to an old problem.

(c) Hobbs as a theorist of the Earth

The quality of Hobbs's insights into the structure and history of the Earth varies almost exactly in inverse ratio to the magnitude of the problem he was tackling. When he is discussing large geophysical and astronomical problems such as the rotation of the Earth on its axis, or the source of the tides, his own provincial ignorance of recent work, and gaucheness at handling complicated—and often mathematical—notions, and his inability to grasp objections to his own views, stand out most noticeably. When dealing with large issues he regularly oversimplifies and begs questions. (Though this is not to deny that he has some sharp and pertinent criticisms to offer of other contemporary theorists of the Earth such as Burnet and Woodward.)

Furthermore, even when dealing with more immediate, concrete realities such as the strata, rocks, fossils, and the like, Hobbs is sometimes hindered by being ignorant of published work publicly available. For instance, he too readily assumes that all rocks are to be found stratified, or that the remains of *terrestrial* creatures are not to be found in a properly fossilized condition. And he is apt to ignore or oversimplify issues which greater naturalists—such as John Ray—had contemplated for years. He could, for instance, simply assume without argument that fossils were organic remains; or dismiss peremptorily all other theories of the origin of rivers than that they took their origin from rainwater—both issues being ones which had genuinely taxed scientific minds in his own time. One has the impression of a mind unused and impervious to criticism, creating a system which was suspiciously self-confirming, because it had never had to think probabilistically about problematic evidence.

Likewise, it is *prima facie* strange that Hobbs makes no explicit use of natural history data potentially available to him in the form of articles in the *Philosophical Transactions of the Royal Society*, and the natural histories of Plot (1640–1696), Lhwyd (1660–1709), Leigh (1662–1701), Ray etc. It is not clear in most cases whether he read such authors or not or whether he believed his own methodology of strict observation and induction required him to desist from making use of the observations of others.

All in all, these habits lent to Hobbs's theory a rather assertive and one-dimensional character. He set out his beliefs in the form of postulata. He rarely supported geological references with detailed local evidence. His diagrams (sadly lost) must have been abstract and idealized. He made no attempt to meet hostile criticism or to deal with anomalous cases.

Yet, where Hobbs had observed for himself, he appears both as a sharp observer, and as an intelligent interpreter of the significance of phenomena he had seen. He claimed to have viewed strata in half the counties of England, though he mentioned by name only Dorset, Devon and Cornwall, and was vague about particular geographical locations. He was probably exceptional rather than typical in his generation in being a naturalist who was both highly familiar with the structure, distribution and composition of rocks in a particular locality, and also concerned to generalize that knowledge into a theory of the Earth, whose universal validity would explain local phenomena.

Hobbs's geological intelligence is apparent in the following areas: firstly, for Hobbs, the key fact, and the key explicandum, about the terrestial parts of the globe is that they are stratified. The fact of stratification, and the need to explain it, is the hub around which all other aspects of his ideas about the Earth revolve. This is itself highly interesting, since observation and analysis of the *strata* comprised only a rather minor part of contemporary theories of the Earth and of local natural histories. In this prespect, Hobbs's interest in the strata is more typical of geology two or three generations later, rather than of his own time.

Hobbs's way of handling the strata is also distinctive. Unlike a theorist such as Woodward, he was not concerned to show that the strata are universally found in the same situation. Unlike natural historians like Plot, Leigh, and Morton, he does not concern himself with compiling lists of the order of the strata in particular geographical locations, naming the particular rocks. He has none of the ambition of later geologists like William Smith (1769–1839) to establish the general order of the strata, and to trace it across the British Isles.

Rather, Hobbs's concerns were twofold. Firstly, to establish the range of typical characteristic *positions* of strata. That is to say, under what circumstances they were to be found parallel to the horizon, or parallel to the line of the countryside, or in no apparent order at all; or where they were broken etc. And then secondly to establish—to deduce, in a way which reminds one of the thought of Hutton—what this necessarily demonstrated about the agency which had created the strata in the first place, and brought them to their current positions.

In other words, he was concerned with forces, pressures of a 'tectonic' nature. Thus for Hobbs, in a way not at all characteristic of contemporary geology, strata were the key to Earth history (*cf.* for example, his highly acute remarks on the significance of the phenomenon of stratification as such as a clue to Earth history).

Almost alone in his generation, Hobbs grasped clearly that the very fact that strata existed (*i.e.*, that rocks were found bedded, and that they contained the remains of organic creatures) demonstrated beyond dispute a marine origin for them. And the fact that the normal condition of strata was to lie horizontally demonstrated that natural, gradual, and relatively calm processes had placed them there—not Burnet's crustal collapse, or a Deluge, or Hooke's massive earthquake and volcanic movements. Similarly, the angle at which the strata of hills ranged to the horizon demonstrated the reality of subsequent upthrust from below. Hobbs's tidal origin theory of the strata is similar to those of many French naturalists of the period and the later eighteenth century such as de Maillet, Le Cat and Buffon, all of whom attributed stratification to the action of the tides. Of course, unlike those naturalists, Hobbs had no desire to advance such a theory in order to attack Genesis. But it is nevertheless true that Hobbs's theory of the formation of strata was possible only *because* he had taken such a liberal reading of Genesis, and because he was so committed to the constancy of Nature and to natural causation. He would hear nothing of any cataclysmic theory of the origins of fossils.

Clearly Hobbs also advances his theory of the gradual formation of stratified land by the sea precisely because this is the phenomenon which he had been actually observing gradually occurring on the Dorset coasts—above all in the vicinity of Chesil Bank—for many years of his life. Hobbs shows, in his theory of the formation of strata, a closer integration of extensive personal observation—above all of current processes—with theorizing than is characteristic of more famous theorists of the Earth of his age—men such as Burnet, Whiston, and, perhaps, Woodward. The power of Hobbs as a geological theorist lies in the fact that he lived on the coast. The relations between land and sea, water and matter, are crucial to all good theories of the Earth.

The other related feature of Hobbs's analysis of the Earth which was important and pioneering was his construction of a number of sections of the strata (though clearly it is difficult to write about them, they being now missing). Not only were they the first and most extensive series of sections accompanying a late seventeenth century or early eighteenth century theory; but in type they seem to have been of a kind not commonly found again till the nineteenth century. For they were not sections of particular locations, as one finds in Strachey and Whitehurst (1713–1788), but rather generalized ideal types, highly stylized and conceptualized, illustrative of typical situations (possibly a little like the block diagrams of faults which Farey inserted in his *General view of the agriculture and minerals of Derbyshire*). They embodied and illustrated a theory of how the strata had been formed. Such a graphic technique was not absolutely new—Steno's diagrams[17] are of very much the same kind (though Hobbs seems to owe nothing to Steno). But it is for that reason of no less interest as a sign of how a man of some intellect and much observation had been able—possessed as Derham said, of 'a clear head'—to conceptualize the typical and exceptional in terms of a significant history.

To place greatest stress on Hobbs's analysis of the strata while relatively neglecting his theory of the pulsation of the tides due to the inner heart of the earth or his astronomy, is not simply to select for attention and praise the most 'progressive' parts of his work; nor is it to fall victim to modern disciplinary boundaries which would separate Hobbs's 'geology' from his 'astronomy', his 'oceanography' and so forth. It is rather to point out that on Hobbs's own statement of his intentions, the problem of the strata and of the fossils they contained was his chief concern. After all, his own title page announces: *The Earth generated and anatomized, wherein is shewn what the Chaos was; how and when the Oyster-shells, cockle-shells and all other Marine productions were brought upon, and incorporated in the rocks and mountains of the Earth.*

For this, his natural philosophy of generation was to be invoked as a general framework within which his specific mechanisms for explaining strata made sense, and his discussion of the tides was a necessary digression. Moreover, when he came to reorganize the paper he had submitted to the Royal Society into his treatise, further reflection clearly caused him to decide to place the problem of the strata, and the formation of the present condition of the Earth—problems which had been relatively unimportant in his Royal Society paper—to the forefront, as the organizing principle which was the intellectual thread running through all his work. And, further, it was within his discussion of formation of the Earth that Hobbs, confessedly an unlettered man, could speak from most experience and with greatest interest.

Hobbs was swimming against the tide in his day (as the quick dismissal of this theories by the Royal Society clearly shows). He was putting forward a philosophy of an active, living Nature at a time when Newtonianism was stressing *God's* activity, and the passivity of Matter itself. He was offering a liberal view of the Biblical creation at a time which piety required more orthodox readings. He was seeing Nature in terms of process, through alchemical philosophies, at a time when the mechanical philosophy was replacing the chemical philosophy. It is not surprising that Hobbs's paper was virtually neglected by the Royal Society. Had *The earth generated and anatomized* been published it would probably have received rather less attention than, say, John Hutchinson's (1674–1737) anti-Newtonian *Moses's Principia*.

Yet there is much in his natural philosophy which is an indispensable background for his observations and technical mechanisms, and much in his specific notions which back up his philosophy. Not until new natural philosophies came into vogue associated with men such as Hutton, which were not precisely like Hobbs's, but which carried many of the same burdens—the stress upon the activity, and organization of the Earth, upon its being transformed gradually and continuously, the demotion of Biblical literalism and of miraculous agency—could the study of the Earth actually rediscover many of the important foci of Hobbs's interest.

The manuscript of *The earth generated and anatomized*, with some notes on editorial practice

The manuscript of Hobbs's treatise from which I have worked appears to be the only one in existence. It has never before been printed. It was purchased by the British Museum (Natural History) from Messrs Dawson's of Pall Mall who had advertised it for sale in their catalogue no. 240 (1973). The history of the manuscript is obscure. Certainly the William Hobbs will of 1743 (p. 8), which I take to be the will of the son of the author, makes no mention of any scientific papers. Written on the vellum cover of the manuscript are the words 'Hobbs Animadversiens' which suggest that at some stage it fell into French hands. On the first sheet of the manuscript are the words:

J. Bailey
August 1825
I bought this MS. in the borough of Southwark.
Shortly after I had made the purchase, mention was made of it, the Title quoted partially,
&c. in the Times newspaper.

J.B.

The manuscript is a folio of seventy-five sheets, bound in contemporary limp vellum. It measures 33 cm × 21½ cm. The text comprises three main sections:

 (*a*) To The Reader (unpaginated)
 (*b*) The main body of the text (paginated 1–54, recto only)
 (*c*) Postscript (paginated 1–18 recto and verso).

It seems to have been written in two hands. The first two thirds or so of the main treatise is in a neat, polished, mature hand, and considerable care was obviously taken in the production of the manuscript. Many passages of the rest of the treatise and the Postscript are also in this hand. This hand can be identified with confidence as that of the William Hobbs who wrote and signed the two letters to the Royal Society in 1709 (though these two letters were penned less formally). It is perhaps not surprising that Hobbs, a sometime excise officer, could write a handsome hand.

The rest of the treatise and the Postscript are in a different hand, which appears less mature, less confident, less fluent and less regular. It would seem as though Hobbs obtained the help of a second hand in aiding him to complete the copying out of the manuscript—possibly a member of his family (this second hand may well be that of a juvenile). This might indicate that Hobbs showed a certain impatience during the later stages to complete the work. In many places the two hands succeed each other every few paragraphs. Errors in the second hand are often corrected by Hobbs's original hand.

Hobbs signed his manuscript, at the end of the section 'To The Reader'. His name has been scratched out and almost obliterated. The signature—so far as it still appears—seems to be in the same hand as the earlier part of the treatise, and rather different in style from the signatures as appear on Hobbs's letters to the Royal Society. It is not clear whether Hobbs himself scratched out his name, or whether that was done subsequently, or, indeed, why this was done.

At no other place in the manuscript did Hobbs sign his name. The title page refers to the author as 'W.H.'. Identifying books by initials was, of course, common practice in Hobbs's day. It is not clear why Hobbs chose to appear simply as W.H. It is conceivable that he did it to avoid all connexion with Thomas Hobbes. It is rather more likely that, as an unknown author, he might expect more notice to be taken of his tract if it were virtually anonymous than if known to be by an author who had no other work in print. This assumes that Hobbs intended his treatise to be printed, or at least circulated.

The manuscript bears one date: 'July 1715'. This is at the end of the section 'To The Reader'. The numerals '15' have been scratched out. Since we have no rough drafts of the treatise it is difficult to base a reconstruction of the chronology of the composition and penning of the treatise upon more than plausible guesswork. Hobbs had presumably been making observations and having ideas about the structure of the Earth from the late 1670s, since he wrote in his letters to the Royal Society that his experience in these matters was of more than thirty years.

It seems to me most probable that Hobbs first set out the major ideas of his treatise soon after he completed his register of the tides, *i.e.*, from about 1704 onwards (assuming that he was dismissed from the excise service at about this time, this might have given him the requisite leisure). Presumably, as he himself claimed, he had kept his tide record in order to confirm a hypothesis about the independence of the tides from the Moon; and having demonstrated that, he would wish to write up such an interpretation. He then seems to have taken no further steps, until he made contact with the Royal Society in 1709, sending them two letters (at least), the first of which spelt out his arguments about Motion and the Moon, and including his 'An essay concerning Motion', which set out many of the ideas (*e.g.*, about trade winds) to be found in *The earth generated*.

Why Hobbs decided, after his 'snub' from the Royal Society, to write up his ideas into the form in which they at present stand, is unclear. It may have happened exactly as Hobbs himself states in his 'Postscript', which I assume was penned with the rest of the treatise in 1715. At the beginning of the 'Postscript' Hobbs wrote: ''Tis almost ten years Since I composed what is generally contained in the foregoing Treatise', indicating perhaps that that was written about 1706, and that he had put the work aside because he had despaired of being able to demonstrate the global movements and times of the tides. However, as he went on to say, he had recently come across a work, the *Lightning column*, which solved such problems for him, and opened the way forward. Having read the *Lightning column*, Hobbs presumably felt in a position to expound his chief ideas in the body of his treatise, confirming them from deductions from the global pattern of the tides in the 'Postscript'.

Comparison between the papers Hobbs sent to the Royal Society in 1709 and *The earth generated* suggests a considerable amount of reworking in the interim. Certain areas of Hobbs's thought appear much more strongly in the later treatise (*e.g.*, his alchemical ideas), and what had been separate and isolated sectors of his theories (*e.g.*, his interpretations of Monsoon winds) had by 1715 been worked into a comprehensive philosophy of the globe. One assumes that had the manuscript of *The earth generated* been in existence, in a reasonably complete condition, in 1709, he would have sent that to the Royal Society.

A Note on editorial practice

In rendering Hobbs's manuscript into print, I have tried to strike a compromise between two desiderata: (*a*) the aim of providing an exact transcript of the manuscript, (*b*) the aim of presenting an easily read version. To the first end, I have retained the spelling, capitalization and punctuation of the manuscript. I have also indicated deletions, alterations and other emendations in the manuscript, where these might have some intellectual significance, rather than being mere slips of the pen. I have tried to do this inconspicuously, distracting the eye as little as possible from the flow of the line. At the same time I have chosen not to try to reproduce every idiosyncrasy of the manuscript, since that would have been to hinder general readability. Any reader wishing to hunt among such idiosyncrasies for further clues in the appearance of the manuscript may consult it on the permission of the Librarian, in the Palaeontology Library, of the British Museum (Natural History). On a number of occasions, Hobbs deleted a word or phrase and substituted one nearly identical. I have chosen not to burden the present text with these variants, since in my judgment the variations are of no intellectual significance, and I have simply transcribed Hobbs's final intention.

Hobbs had a highly elaborate practice of representing relative degrees of emphasis for different words and phrases by the use of distinctive sizes and thicknesses of script and different degrees of italic and gothic script. In a rather simplified manner I have tried to retain Hobbs's emphases, largely by the use of bold type. I have used square brackets for the occasional [sic] and to mark the page numbers of the original manuscript. All other uses of square and round brackets are Hobbs's.

An impressive feature of Hobbs's original manuscript must have been the diagrams—some twelve in all, probably contained on about five or six sheets. Of these only one sheet survives, containing four diagrams (nos 9, 10, 11, 12). These have been reproduced here. For the convenience of the reader I have attempted to reconstruct the remaining diagrams. I wish to thank Dr S. Conway Morris for his great help in this task.

Hobbs's scientific drawings were clearly prized in his day. He had sent similar drawings and models to the Royal Society in 1709, including a map of the Isle of Portland. William Derham remarked on their quality. At least some of these survived through to the mid-eighteenth century, for notice of them appears in manuscript catalogues of the collections possessed by the Royal Society ('An inventory of the subjects of natural history in the Repository of the Royal Society, Nov. 21, 1763' contains a reference to 'Representations of the Earth's Strata; 4 pieces, and also of the Bell fish in a box' which are clearly Hobbs's). When or how they were finally lost is not known.

Notes

1 Original in Royal Society Archives, MS Ex 1. 13; copy in Letter Book Supplement, G–H, copy, pp. 357–406.

2 At the end of his 'An Essay concerning Motion', *op. cit.* (note 1).

3 The following section is based chiefly on the surviving minutes of the Excise Board, preserved at the library of the Custom House in London, and to a small extent on the minutes and correspondence of the Customs Board, also preserved there. Since these records date only from the 1690s, it has been impossible to trace the origins of Hobbs's connexions with the excise authorities. It appears from these that William Hobbs, Sr, was already excise officer at Weymouth in 1698, when, for a short period of time, his son substituted for him while he was sick. The younger Hobbs was posted frequently to new 'Rides' in such places at Taunton and Worcester, moving it seems to Hindon in Dorset probably in 1703. Meanwhile his father continued as officer at Weymouth, becoming officer at Bridport in November 1704. In January 1704/5 Hobbs, Jr, was dismissed for dishonest returns. The same fate befell his father in April 1705. In August 1705 one of the Hobbs's—I presume the younger—was reinstated to a Devon collection. He served other Devon collections in the next few years. The last mention of him I have found in the minutes is in 1707. The other Hobbs does not figure in the minutes at all after his dismissal. I am grateful to H.M. Customs and Excise for permission to use and quote from their archives, and in particular to Mr F. R. Blanchard for his courteous help in explaining the archives, and the structure of the excise service to me. See also E. Carson, *The ancient and rightful customs* (London, 1972).

Prima facie there might seem to be some connection between William Hobbs of Weymouth, author of *The earth generated and anatomized*, and the man of the very same name who in 1714 published a tract, *A new discovery for finding the longitude*, printed for the author who was found at the sign of the Porter in Baising Hall Street, London. Apart from their almost exact contemporaneity, both works show interest in broadly similar problems—of navigation at sea and its difficulties. But it is almost certain that these William Hobbs's are in fact different men. The author of the longitude tract was a clockmaker (quite probably the man of that name apprenticed in 1672 in London). There is no intellectual cross-referencing between the two works. The London Hobbs was chiefly interested in solving the longitude problem through fancy pieces of clock work; the Weymouth Hobbs was deeply interested in the much more philosophical problems of the figure and motion of the earth.

4 Weymouth and Melcombe Regis Corporation Records, 1699–1724, no. 259, 185 verso.

5 The Weymouth and Melcombe Regis Corporation Records (Item S 259 Minute Book, 1699–1724, p. 2366, 8th March 1722/23) mention a Mr William Hobbs as a school master. Dr Hugh Torrens has kindly drawn my attention to the fact that Joshua Childrey was in contact in 1669 with a 'friend' from Weymouth who taught 'mariner's art' and was interested in tides. See A. R. Hall and M. B. Hall, *The correspondence of Henry Oldenburg* (Madison, 1969, **6**: 109). It is *just* conceivable that this was the elder Hobbs, but there is no need to suppose this, since the teaching of mariner's art must have been an important and integral part of the economy of a town such as Weymouth. In 1753 a school was set up for training boys who would go to sea; see M. Weinstock *Old Dorset* (Newton Abbot, 1967).

6 The will is at the Dorset County Record Office, Dorset Archdeaconry Records, DA/W 1743/32.

7 *Op. cit.* (note 1).

8 Royal Society Journal Book, X 1702–14, p. 215.

9 Royal Society Register Book, IX, pp. 268–73.

10 Royal Society Journal Book, X, 1702–14, 25th May, 1709. I assume that on both occasions 'his' in Woodward's sentence refer to Hobbs not to Woodward.

11 Royal Society Journal Book, X, 1702–14, 13th July 1709.

12 This letter is Sloane Mss, 4042, f. 13, in the Department of Manuscripts, British Library. There is a copy in the Royal Society archives, LBC 14 276. The Bear Inn was in St Mary's Street.

13 A similar provincial, contemporary with Hobbs, was Conyers Purshall, a country physician (? at Bromsgrove), who wrote *An essay at the mechanism of the macrocosm* (London, 1705). Purshall complained of his isolation from learned society. His book shows a mixture, similar to Hobbs's, of rather acute criticism of fashionable mechanical and Newtonian philosophy, with some extremely eccentric and old-fashioned views (*e.g.*, an apparent denial of Copernicanism). Like Hobbs, Purshall believed that the pressure of the aether accounted for all heavenly motions, but there the similarity ends.

14 See D. E. Allen, *The naturalist in Britain* (London, 1976), for an account of the social development of the natural history sciences in Britain.

15   This account is derived from sources such as J. Coker, *A survey of Dorsetshire, containing the antiquities and natural history of that county* (London, 1732); John Hutchins, *The history and antiquities of the county of Dorset* (2 vols, London, 1774); H. J. Moule, *Old Dorset* (London, 1893); T. Perkins and H. Pentin, *Memorials of old Dorset* (London, 1907); and G. A. Ellis, *The history and antiquities of the borough and town of Weymouth and Melcombe Regis* (Weymouth, 1829). For a rich guide to the literature on Dorset see R. Douch, *A handbook of local history: Dorset* (University of Bristol, Bristol, 1952, revised and corrected ed., 1962).

16   For a good bibliography of books and articles on Dorset geology see G. M. Davies, *The Dorset coast, a geological guide* (London, 1935: 113–22).

17   M. J. S. Rudwick, 'The emergence of a visual language for geological science', *History of science* **14** 1976: 149–95.

The earth generated and anatomized:

Text

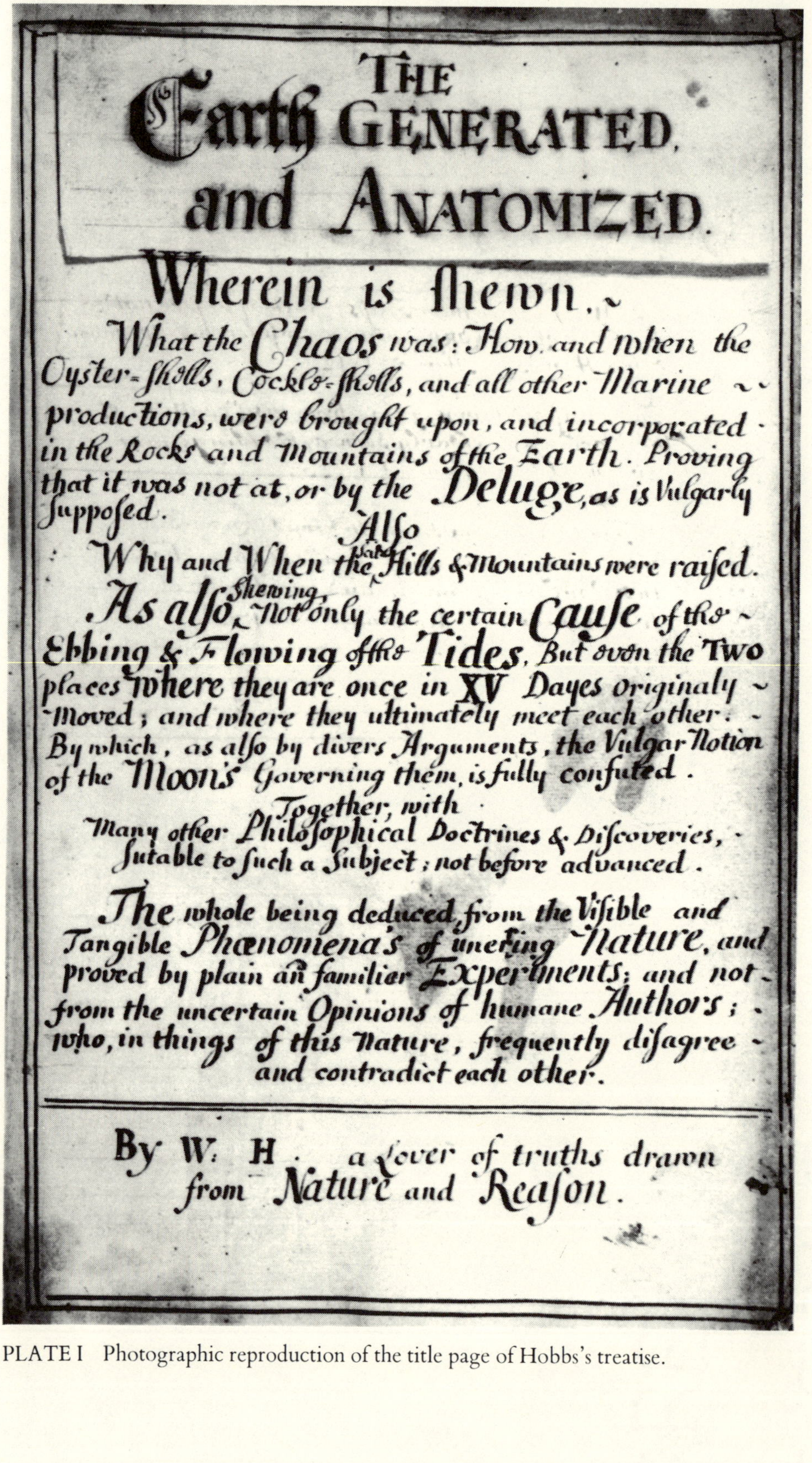

THE Earth GENERATED, and ANATOMIZED.

Wherein is shewn,

What the Chaos was: How and when the Oyster-shells, Cockle-shells, and all other Marine productions, were brought upon, and incorporated in the Rocks and Mountains of the Earth. Proving that it was not at, or by the Deluge, as is Vulgarly supposed.

Also

Why and When the Hills & Mountains were raised.

As also, shewing not only the certain Cause of the Ebbing & Flowing of the Tides, But even the Two places where they are once in XV Dayes originaly Moved; and where they ultimately meet each other. By which, as also by divers Arguments, the Vulgar Notion of the Moon's Governing them, is fully confuted.

Together, with

Many other Philosophical Doctrines & Discoveries, sutable to such a Subject; not before advanced.

The whole being deduced from the Visible and Tangible Phænomena's of unerring Nature, and proved by plain an familier Experiments; and not from the uncertain Opinions of humane Authors; who, in things of this Nature, frequently disagree and contradict each other.

By W. H. a lover of truths drawn from Nature and Reason.

PLATE I Photographic reproduction of the title page of Hobbs's treatise.

THE EARTH GENERATED AND ANATOMIZED

Wherein is shewn
What the Chaos was: How and when the
Oyster-Shells, Cockle-Shells, and all other Marine
productions, were brought upon, and incorporated
in the Rocks and Mountains of the Earth. Proving
that it was not at, or by the Deluge, as is Vulgarly
Supposed.

Also
Why and When the Said Hills and Mountains were raised.
As also shewing not only the certain Cause of the
Ebbing and Flowing of the Tides, But even the Two
places where they are once in XV Dayes originaly
Moved; and where they ultimately meet each other.
By which, as also by diverse Arguments, the Vulgar Notion
of the Moon's Governing them, is fully confuted.

Together with
Many other Philosophical Doctrines and Discoveries;
Suitable to such a Subject; not before advanced.
The whole being deduced from the Visible and
Tangible Phænomena's of unerring Nature, and
proved by plain and familiar Experiments; and not
from the uncertain Opinions of humane Authors;
who, in things of this Nature, frequently disagree
and contradict each other.

By   W.   H.   a Lover of truths drawn
from Nature and Reason

—Wine are in their Feasts; But they regard not
the works of the Lord; neither consider the operation
of his Hands. Isai 5 v. 12.

The Invisible things of God from the Creation of
the World, are clearly seen by the things that
are made. Rom. 1 v. 20.

For he hath Given me the true knowledge
of the things that are; So that I know how
the World was made, and the Power of
the Elements. Wisdome 7 v. 17[1]

To the Reader

It may be discovered by the Title Page, That the ensuing Lines, do not only differ, from the opinion of many Learned Writers, of This and the former Ages; but alsoe assert other Propositions, not before advanced: And though this may be a presumption in the Author, yet if it be duly considered, 'twill apear highly needfull to have been done, for many Ages past; For tho' the Mathematicall Sciences, are Treated of, with Such Nicity and unanimity; as that there seems no Room to correct what has been Written; not but Little for any further Advancement; Yet in those Sciences, 'tis quite otherwise; For there are Such Diversity of Opinnions concerning, even the Visible Appearances in Nature; that a man knows not which is true, or which is false: From the Consideration whereof, I was induced to Lay those opinions, wholy aside; Not for want of Gratitude to Some and due Respect to the Rest of those Authors: But to try what I could Discover, as well by observing the Said Phænomena's; as by Searching into the Causes thereof; in hopes that Some Right Foundation might be Laid for *this*, as well as other Sciences.[2] In Which Enquiry, haveing had the advantage of makeing my observations in near half the Countys of this Kingdom, as well in the Inland parts, as on the Sea Coasts[3] I allways found; That in all Deep *Roads*, *Pitts*, *Quarries*, *Mines*; etc. the Constant Position of the Stratas, or rather Beds of Earth; whether it was Level, Hilly, or Mountainious; were exactly parrallel to each other; as in the 1st: 3^{d} and 4:th Fig:[s4] is described: Which Position I never before found treated of, tho' it Teacheth, as much or more, than any Terrestriall Phænomenon whatever.[5]

After this I observed the Position of the fish=Shells[6] and found them to be immassed, as well in the Solid Rocks & Stones, as in earth, and looser matter, and tho' they are not generally So,—in all places,—yet what is described in the following Figures, is no more than I have actually seen, especially on the Sea Coasts, for Severall Furlongs; and Sometimes for Miles together; and in Some places, even to the depth of 2 or 300 foot perpendicular;[7] So that you may certainly depend, upon what you [p. 2] See therein decyphered. Now when I had made these observations; I further observed, That if a greater or Lesser Tree (or any of their branches) be Sawn a Sunder; At the Ends thereof, the Veins will appear as exactly parrallel as they do Circular; and well knowing, that the Generateing, the Sphericall Earth; was the Same, in its Kind, with any of these cilindricall Vegetables; I was thereby taught, that the Beds of Stone Cley, earth, &c. were at first, not only as is Said parrellel, and Circular, but also Levell, and Horizontally posited: as in Fig: 8th is described. Whence I plainly discovered, That the shells, were then; to wit, when the matter of the Earth was plain, Soft, and Levell and no Hills raised, brought upon the Earth; and incorporated, and mingled therewith: And that when the Earth was fittly hardened, the Hills were thrust up; together with the Shells and other Marine productions that were then incorporated, and are now remaining therein. And this we may be assured, was the real Means whereby they became **Posited**, (as we frequently find them,) so many hundred feet higher than the Ocean, in which their Species do now reside. All which is evidenced, by their very Position in the Rocks; as is described in Fig: 2:^{d}.[8] Because the Matter must be both soft and level when they were thereinto admitted. For every Housewife can testifie, that the Fruit, must be put into the Cake or puding before, and not after it is baked or hardened:[9] And as the Earth was then undeniably soft and tender; so 'tis as evident that it was level and that there could not, then be any Hills or Mountains raised; For if they Should in that State, the Soft Matter Would Subside, as fast as it was Elevated; as dayly Experience teacheth and Confirmeth.

Now having delineated and transcribed these Visible Symbollicall Characters, imprinted by nature in the Earth, and incerted the Inferences, that did naturally flow from them, upon the following Sheets, I found them to differ from other opinions, as aforesaid: Whereupon I presumed to Examine them by that admirable Compendium of the Creation, Contained in the two first Chapters, of the Book of Genesis; wherein We have a more true, and profound Account thereof; than in all the philosophical Books yet extant; And haveing thus done, I found my mean Endeavours, were thereby So far Justified; [p. 3] as that they seemed to be, but as a Naturall Comment, or Paraphrase upon the Said Chapters; And that this may the better appear, we will here incert an Abstract thereof: In which it is to be noted; That When Moses had given an Account, of the Creating the first Matter, out of Nothing; He then intimates or tell us—[10]

First, That it was a deformed Mass of Water; and that the Spirrit of God moveing thereupon produced Light.*

" 2.dly That God Seperated the waters designed for the Earth from those of which "the heavens were made.

" 3.dly That after the Waters were gathered together, the dry Land was Raised up, or "appeared above the Waters. Whereupon the Earth broght forth Grass &c. "abundantly.

4:^{thly} He formed or made the Sun, moon, and Stars.

5.^{ly} He commanded the waters to Bring forth Fish, and Fowles and

6^{ly} "He commanded the Earth to bring forth Horses, Cows, Sheep Lyons, Dogs, &c. and all other Living Creatures Whatsoever[11]

And when he had given us this Brief Account of WHAT, was done, he then in the 4.th Verse of the 2^{d} Chapter[12] tells us the Maner HOW, or by what Means it was Effected. (namely)

---

* Note this Light was doubtless the Same in kind (or to which he aludes) with that we now finde on the Surface of the Ocean; for, Especially at Some Seasons of the year, and in hot countries, if a Vessell of salt Water be carried into any Close Roome, in the darkest Night, it will, if stirred, produce Such a light as that I have Seen an hair, by that Light only.

"That the Heavens, and the Earth, were made, by GENERATION: In which Emphaticall, and comprehensive Word, We have the Whole Process of Nature, in forming the Universe, fully described; So that if we know, but what it is to GENERATE, or ANIMATE, we may know the manner how, all Naturall Bodies, from the first Matter, were formed, and produced: And whether I have pursued the Foot=steps, or process, of GENERATEING in the following Lines, I leave to be determined by the Judicious Reader.

[p. 4] But notwithstanding the Excellency of this Account it is to be observed That it has not past unsensured; for some (haply not from their knowledge in the Works of Nature, but for Reasons best known to themselves) have objected, That the Space of one day, was too Little, for makeing the Heavens, when the Earth required 5: Others that the Order in which they are Said to be made, was Sometimes inverted; as in the 4th Verse The Heavens are Said to be made, before the Earth; and in the Close of the Same, the Earth is said to be made before the Heavens; so likewise in the first Words, 'tis said the Heavens and the Earth were Created; and yet after that, we have an Account of the Makeing them both: so also in the Middle of the Week (to Wit) on the 4th day, 'tis observed that the Heavens were made; during which some would infer, That the forming of the Earth was laid by; and reasumed the day following &c.

Now that these seeming Contradictions may be reconciled, I shall presume to Shew, that there can be no Methodicall account given of the Process of Nature in Generateing or makeing the universe, or of that part called the Terrestriall Globe, or any Generated, or Animated Body Whatsoever.[13] 2dly That the pretended Contradictions; are no other, than Such, as in all probability, the Authur designed; and what Nature, Reason, and our dayly Observations, do allow; or may be justly reconciled unto. And to explycate the Matter; Lett us Suppose that a Statuarist should carve a Figure, (e.g) of an Apple Tree or another person, should make such a Machine as a watch; of which let the first, (as it resembles), be compared to a Vegetable; and the Latter to an Animall: Now tho' these Bodies, when finished, may bear Some Resemblance in Respect of the form, or Regularity in composure; or in Respect of Power in moveing &c: to the Bodies with which they are Compared; yet the Order or **Method** in forming, or composing them, is far different, from the Gradations or Proceedings of Nature in forming Naturall Bodies: For when the Artist, in one day, is giving Shape, Suppose to the Body; or but to any Leaf of the Tree; during that time, all the rest of the designed Leaves and Fruit [p. 5] will remaine formeless and wholy unfashioned; and so of the Severall parts of the Watch or other Machine. But in the process of Nature 'tis quite otherwise; For 'tis well known, That no part of any generated Body is at a Stand, or unaffected, whilst the other Parts, are composing: But that the whole, and all its Parts, are carryed on, by a common Coagitation, and Generall Gradation: So that 'tis impossible to say, this day the Foot of the Head; and that day the Bones &c: in any Fœtus, was formed, or that any part or Member thereof was perfected; before the whole is fully Animated: (And the Same may be understood of a vegetable) Which being well-known to this Wise and Learned Author, he accordingly treated of the Creation promiscuously; by inverting and counterchanging the Order or Method thereof; as is before observed: In which he so Signalised his Wisdome, that instead of being censured or exploded, he highly deserves the greater Admiration.[14]

Now Seeing 'tis Evident, that the Heavens and the Earth were Generated, and that therefore, there could be no Methodicall account Given, of the perfect order in which they were made: We may from these considerations, justly conclude, That the Process of Nature, in Generateing that Part of the Universe called the Heavens; was as certainly carryed on, in the 1st 2^d and 3^d Days (to wit) whilst the Earth was forming; as it was on the 4th Day; notwithstanding they are not till then, said to be made: And also That the Waters, did as Certainly bring forth all Sorts of fish, if not Fowles, (at least) as well on the 3^d and 4th Days; as they did, both fish, and Fowles on the 5th Day. For it appears, by the foregoing Abstract that as soon as the dry Land, was raised above the Waters; it immediately became fruitfull and most plentifully brought forth Grass, and other Vegetables. With what Shadow of Reason then can it be supposed, that the Waters, which had their Being from the first Beginning of the Creation, should be fruitless and barren untill the 5th Day (to wit) 'till after the Heavens and the Earth were made?[15] For 'tis but naturall to conclude That at the same instant in which matter was Created or produced, it was likewise elemented and imprincipled with its productive Nature, or plastick Qualification: It being absurd to Suppose, That the said matter or either of the Elements, could Exist before, or without, its essentiall properties. And this the Father of Solomon fully affirmes when he tells us, That by the Very word of the Lord, the Heavens and all their host were made psal: 33. v. 6^{16} (i.e) Matter was thereby Created, and [p. 6] imprincipled in order to generate them, as aforesaid. So that in all probability the designe of Moses, was not so much to ascertaine, an exact Mechanick, or literall Order, for makeing the Heavens, and the Earth; as to convince and Assure us, That they were made, & Not Eternall: And therefore what I have said in the following lines concerning the Shell=Fish &c. being brought forth, or generated before the Fifth day; (to wit) before the Hills were raised, (which some perhaps may object against,) is not so contrary to the said Compendium but that it is comformable to the designe of its Author; And is fully warranted, from the process of Nature, in Generateing or Animateing of Matter; and also abundantly confirmed by the Shells being Visibly found incorporated in the Rocks as aforesaid. And what is more by this Hypothosis, Nature, Reason, and our dayly Observations, will be mutually reconciled; and the Manifold Absurdities, so inevitably arising from the contrary; be dessipated & prevented.

And forasmuch as diverse other things contained in the following lines are either New, or at least but rarely treated of; I am not unsensible, That there are Two things wanting, which might recommend it to the Readers approbation; (Namely) First, That it proceeds not from the Pen of Some famous Learned Author: And Secondly That it is not confirmed by the Concurrent Testimonies of Such.

As to the first of these, I readily grant, that if it had been promulgated by the Learned Grotius, or DesCartes, or by our adjacent neighbour [two words undecyphered] I mean the Author of the famous or rather infamous Leviathan their very names might have given it such a Sanction as would perhaps Silence Objections But insomuch as I am not endowed with the learning of the former, so I think fitt for some Reasons to acknowledge that notwithstanding the Vicinity before mentioned I am not any way related to the latter. I must therefore commit it, as it is, destitute of the said advantages to the Censures of the Judicious Reader.[17]

But however to Shew, that it ought not to be rejected, barely for want of these accomplishments, in the Author; I shall endeavour to prove, That Learning especially of Languages;[18] conduceth but little, to enquiries of the Nature: For 'tis evident That [p. 7] Learning (I speak not to lessen its Excellency, for 'tis what I highly honour and admire:) may teach us what is *already* known [to others;] But 'tis not that alone, that will finde out *new* Discoveries; or infallibly unfold the deep misteries, imprinted in the Volumes of Nature. For know in Musick, he that learns the Theoretic Part of *Contra Puntum* can play but only what is first composed, Whereas he that knows the latter can perform what the former never played,[19] though he that imploys his time, in Reading what is already published; may know what other men have known; Yet 'tis the Setting one thing against another, by *Contemplation*, that finds out what his Authors never knew. Had what I have herein after treated of, or discovered, depended only upon **Mathematicall Sciences**, or **Orientall Languages**; it had not been a Secret so long, as 'till the Days of *Euclid*, or *Aristotle*; But as it is derived, from a Dilligent Searching into *Natures* **Symbolicall Characters**, the bare knowledge of a multitude of Words, can contribute but little thereunto. How absurd would it be to thinke That Columbus's dark and Midnight conceptions of the Western World, that had lain so many Thousand Years in Obscurity, and undiscovered, Should proceed from Learning Spanish, Welch, or Irish; Why then Should we undervalue our selves or Language, so as to thinke the knowledge of Greek, or Hebrew; as the Vulgar doe, to whom I am now Speaking, will teach us the Secret Misteries of Art or Nature? Is every Native of Greece, a Philosopher by Virtue of his Language? or every Jew, a Mathematician by means of his? Nay I'll appeal to Reason, Whether the more a man Spend's his time in attaining Such knowledge, he be not thereby the more hindred and diverted, from Contemplation? by which, as is said, all new discoveries are attained. And that they are so, will appear by the following instances; For

Was it not by Contemplation, that Pythagoras found out the 47.th Problem in the I Book of Euclide's Elements.[20] at the Discovery of which so many Oxen were slain in Sacrifice; and on which so great a part of Mathematicall Learning depends. 'Twas [p. 8] also by thinking and not only by Reading, (for whatsoever is prædicated or published, is already known,) that the Reverend Bishop, in the age of ignorance, lost his Life; for Asserting, The *Earth did not Rest on the Sky beneath them*.[21] And 'twas by the same, that the aforesaid *Columbus*, found out that Vast Continent of *America*, for those who after he returned, most ungratefully Slighted, and afronted him. 'Twas also by thinking, That *Archimedes* found how to Discover, the fraudulent compounding of Mettles;[22] at which he was transported as to forgett his Cloaths when Naked. And of later Ages, 'twas that by which the Hon.ble L.d *Napier*[23] of North=Brittain, by his profound and famous Invention of Logarithms, converted Multiplication and Division, into Addition & Subtraction. And, Pardon my presumption, 'Twas also by thinking that I obtained the Discoveries; with which you are presented, in the following Lines. In a Word, Learning produceth Knowledge; Knowledge fitts man for Contemplation; and Contemplation, finds out new Discoveries; From whence Learning was primarily educed; and is still improved and increased. To conclude therefore seeing what is herein advanced was principally obtained by Contemplation; it may nevertheless be true, though it does not proceed from such an Author, as these before mentioned.

2ly   The next thing that's wanting to recommend what is herein Writen; is because, it is not confirmed by the concurrent Testimonies, of such Authors; Which indeed I have very rarely mentioned: But my reasons for so omitting, are First because they could not be obtained: And secondly, Because they are either Needless or useless and,

First, That they could not be obtained, is manifest; Because a great Part of What's herein Advanced, was not before discovered:[24] As the Time When, and manner **how**, and places **where**, the Waters are originially moved; with Divers other things of like Nature; as in the ensueing treatise will more fully appear. And being, as is said, not before Discovered, 'tis therefore impossible to obtain concurrent Testimonies for their Confirmation. And truly had I not been Assured of their Verity, as well as of [p. 9] their Novelty, I would not have given my self the trouble of Writing, Nor you of Reading, what is herein contained; But would even now tho' finished, bury it in Obscurity. Seeing therefore you finde me wholy averse to publish what has been already, *rightly* treated of by others; I hope my endeavours herein will be the more acceptable; and the rather because there are so many that have Writtin, even on the uncontreverted Mathematicall Science who might, (as Mr. Lilly[25] has done, in his treatise on that imaginary and imperfect Art of *Astroligie*) who might I say, have terminated their Labours, with a *Nihil dictum quod non dictum prius*; as haveing left the Science whereon they treated, but in the same state wherein they found it.

And since I have here, as well as in Section the Second, &c. taken Occasion to mention the said Art of *Astrologie*, I must beg leave, tho it be a digression, to show That 'tis even such, as I have Deemed it to be.[26]

And 1st That it is but imaginary is evident, in that it is founded on an imaginary, and invisible Influence, of the Planets and Fixed=Stars; of which, as is Shewn in the said

Section they are wholy destitute; and therefore cannot in the least affect us therewith: Whereas in other Sciences there are real and Visible objects, Whereby to demonstrate, the Rules and Propositions depending thereupon: As for instance, in the Noble Science of *Astronomy*, There are Instruments, and the Visible Bodys of the Stars; by the use, and Observations whereof, their Position, Motion, and Revolutions, are plainly Discovered and computed.

2^{ly} In Geometry and other parts of the Mathematicks, there are real and Visible Lines, Circles and Numbers, to Demonstrate the Powers, Proportions, and affections, that they bear each other. And 3^{ly} in the *Philosophicall* Science now before us, We have the Visible and tangible Phoenomena's in Nature, to guide us in our enquiries, after the Causes thereof; as also for the Inferences and deductions [p. 10] drawn therefrom: But, in Judiciall Astrologie, ther's nothing but Arbitrary Rules, and imaginary Aphorisms, to Guide them in all their Enquiries: So-that if Mr Lilly had thought fitt, he might have changed, even the *Ptolomaick* Table, of the pretended Essential Dignities of the Plannets; by putting *Taurus*, for the day or night house of *Mars*, instead of Aries; for the like house of *Venus*, instead of *Taurus*; and so of the rest, at his pleasure: and accordingly their Effect or Influence, would be equally the same, to the Native or Quærent. For the whole art, unless what Relates to Astronomy, proceeds but from the Arbitrary Rules and Assertions, of their first, or procceedent Authors, as aforesaid.//

And 2dly That the said Art is but *Imperfect*, is evident, Because that in, and near, 66 Degrees of N^o and S^o Lattitude (Viz.^{t}) in *Norway Finland*, and the north part of *Tartaria* and other like places; the Art is as it were wholy Extinct; For if any Native be born, or any Question asked, at, or near the time of the *Suns* (or any other planets,) Setting from the 11th of June to the 13th of December. Or, at the *Suns* (or any Plannets) Rising; for the other halfe of the year; all the signes and Plannets, will be then and there in the First & Seaventh House. and all the rest of the Houses will be void and Tenantless: So that all the Astoligers in Europe, cannot Determine What Plannet, Such a Native, of Such a place is borne under; or what is Lord, or Lady, of the Ascendant; or any other, of their imaginary Houses: By which it appears, that the said Art is imperfect and Deficient; And in those parts of the world; even according to their own pretended Rules, wholy useless: as could be further demonstrated: But I must return.

Haveing already Shewn, That the Concurrent Testimonies of Authors, cannot be obtained, for the Confirmation of *new* Discoveries; I shall now proceed to Shew; that they are Needless, or useless, to confirm, or confute; what is herein [p. 11] Advanced. And first that they are Needless is evident Because, that which Relates to the Discribing, the Phænomena's imprinted in the Rocks, and Mountains, of the Earth; are in them so conspicious, that all persons may Satisfie themselves, of their being Such, and so posited as is described; even by Visibly beholding the Same: And as for that common *Phænomenon* in the waters, I mean the flowing of the *Tides*, it is so generally known and granted, That it would be, not only needless, but even ridiculous, to spend time to conform, that there is such a motion therein. From whence it is manifest, That Such Testimonies, are altogether needless, for confirming the Being, and Position, of the said *Phenomena's*. And as they are therein Needless, so they are also Useless, for confirming the Inferences, and Deductions drawn therefrom: For should we depend upon such Testimonies; to prove what we have assigned, for the *Causes why*; the *manner how*; and the *Time when*; those Characters or Symbols of Nature, were imprinted in the fixed parts of the Earth: Or *how*, & *where*, those in the Marine parts thereof (to wit, the Tides) are originially moved; I say, Should we endeavour to prove, what we have Advanced by Such testimonies; we should have spent time to no Porpose: For as we have already hinted, there are Such Diversity and Variety of Opinnions, concerning many of those Things whereof we have treated; That should they be enumerated, it would Seem to be; *Tot homines, quot Sententiæ*;[27] So that neither of them, how Learned Soever can be depended upon, for confirming or Confuteing what is herein Asserted.

For a Conclusion there=fore, Seeing it is manifest That Such Testimonies, are not only Uncertain; but even needless and useless: And forasmuch as what I have asserted was deduced and Established from the *Visible*, and *Tangible*, Phænomena's of Unerring Nature; My request therefore is; [p. 12] That what I have Written, may be Approved, or Rejected; but as it shall, or shall not be Warranted by, and from, the same Symbolicall Impressions from whence it was Derived; without haveing respect to the Opinion, of any Author whatsoever. And by *Such* a Tryall, or Examination, I doubt not of being Justified, in what is presented in the Following lines:

By yours[28]

July 1715   William Hobbs
                                                             Weymouth

The generating the Earth, &c.[29]
Sect: I.[st]

That fish=shells are found plentifull in the Earth as described in the figures hereunto annexed. That the matter of the Earth was soft and even, when the said shells were admitted thereinto. What the **Chaos**, or first matter was. An Account of the Generateing an animal; and that apply'd to the Animateing the Earth. Why Creatures differ in their shapes, and Why they were not all Orbicular.

WHEN. the Almighty *Creator*, formed Man after his own Image, tho' he did not endow him wth that *perfect Omiscience* which is one of his owne peculiar Attributes, Yet he gave him Such a share of Knowledge, in his innocency; that he was then but a little beneath the Angells of Heaven: And if he had So continued might doubtless, with facility, have given us, an exact Account of the Works of Creation: But his disobediently Aspiring after Knowledge, defaced the Image of his Maker, enstamped upon him, and thereby, clouding his understanding, made his *Ignorance* dayly Increase with his *sin*; so that we have, now, no more knowledge of that great Work, than which God has been pleased to reveal in his secret Writing. Or else what by our investigateing, and prying into the Secrets of *Nature*, we are able thence to Collect: But the Account thereof in the Scriptures, being designed, to let Man know, That as he himself; So all things else,, had their *being*, from that one God which we are commanded to adore; And not to teach us *Phylosophicall* Systems of the Creation; or to fill us with *Metaphisicall* Notions of things Created:[30] the latter way, has, with Various Success, been followed by the most Ingenuous, and Inquisitive Persons of all Ages: Most of whom, tho' their Conceptions have been Deep and Regular; and their Methods and Reasonings, Elegant and Admirable; Yet by their not closely following *Nature's* foot=steps; have therefore Given us but lame Accounts, of its Operations; And more puzled us, with multiplicity of Idea's, than cleared our Understandings:

[p. 14] Now tho' I know my self in divers respects, very unfitt for Such an Undertakeing as has been thus attended; Yet haveing for many Years,[31] diligently Observed, and enquired into the Various *Phænomena's* of Nature; and from thence Collected the following Hypothesis; am the more willing to Communicate the Same; Because 'tis founded and Evidenced, *by* and *from*, such Observations and appearances in nature, as are even *Visible* therein; And therefore Subjected, to the Censure, or Approbation, of the meanest Enquirer: They being chiefly Such as these.

I. First That a multitude of Shells, (to wit,) of *Oysters*, *Musells*, *Cockles*, and severall Marine Productions, both in their Maturity, and some even in their Embrio, are dayly found in almost all the parts of the Earth; not only on, and near its Superficies; but also to the greatest depth we have as yet, been able to arrive: As is evident from the works of many Learned Authors;[32] and also to all Such, as shall please to enquire thereinto, from the said Productions, in the Volums of *Nature*.

II.   Secondly, That these Shells, are not only dispersed in Loose Earth, and the less Solid parts of the Glob; but also in the hardest Rocks, as well of Marble, and Flints, as of other common Stones; and also in Chalk, Copperas=Stones; and all other Terrestriall Fossils;[33] unless Minerals, comõn coals,[34] and Cornish slate, or Tiles; in which I could never yet see any; tho' I have diligently caught them, in the latter, in divers parts, of the South of *Devonshire*, and *Cornwall*; where the Deep Roads, do expose such Rocks to publick View, in many places, to 8. or 10. Foot in Depth, for Severall Miles together; besides the many *Quarries* that are there, to be seen. However, that they are not found therein; is not to be admired; if we consider, how unfitt Such black, and Sulpherious matter was, when fluid, either for the production of such Animals, or for their Sustentation when produced. But tho they are not found in these particular things; yet in the said Stones, and Rocks, and all other matter, before mentioned; They are nevertheless, immassed and incorporated, in such a plentiful manner, as that I have seen some stones, of 50. or 100. Tunns in Measure, to appear almost intirely Shells, or their Impressions:[35] And these lying in severall Beds or ranges of the Earth, to the depth of 60. or 100 Foot: And, on the sea Coast, as well some hundred feet above, as at, and under, the Waters; even for Furlongs, and sometimes, for miles Together. Being there, as well as in the Innland=parts, [p. 15] generally posited as in the following Figures is described. From whence we may inferr, as it cannot be denyed; That the Rocks or whatever contains those Shells, or their Impressions; were, at the time of Receiving them, of a *thin, soft, fluid, substance*; And that the hills and Mountains, wherein those Shells are now found, were not raised, till *after* the said shells were immassed and mingled therein; (see Prov. 8: 24: 25)[36] And this is plain because the earth being *soft* it must therefore *then* be smooth and even; Tho' it is since condenced and hardned into Clay, Stones, Mineralls &c; and the Mountains and hills raised, as we now behold them.

III.   Thirdly, That the body of the Earth, dos now, consist of Severall Beds or Ranges of Clay, Stone, Mineralls &c. Runing alwayes exactly *parrellel* to each other As in Fig. 1.[st][37] The bed or stratum of Clay D.D. is parrallel to that of stone E.E. and that again parrellel to F.F. and soe of the rest, as the Figure, do plainly intimate; which I desire may be well noted, as being a principall Thing herein to be considered.

IIII.   Fourthly That tho' these stratas are *parrellel* to each other, Yet where Hills are raised they are not parallel to the Horizone, but do sometimes cutt the same Angles of 20. or 40. degrees, more or Less as is plainly intimated by Fig: 3.[d][38]

V.   That is they run parellel to one side of the hill, (as the Side A.A. Fig: 3.[d]) then the other side A.B.C. is butted against by the *Ends* of those Beds which are shuting out against the Surface of the Hill, as at A.B.C. aforesaid. And that it is thus Scituate, is *visible* to all persons travelling in any hilly Country, where the Roads are washed into Hollow-wayes; unless perchance, at the time of the Raising such hills; or since, some of them have *accidenttally* fallen from A. towards B. and remaining in such a position have made them seem otherwise, But diligent Enquiry will convince, that their *naturall* Position is exactly parallel as described and accordingly they are generally found.

VI.   That if the Elevation of an hill, is caused by an *equall force* at the same time; thrusting up *all* it's Parts alike; then in such hills [p. 16] the Beds do continue their naturall and primitive Scituation (that is) parrellel to the Horizone, and to each other; and their Ends coming or pointing forth, against the surface, on all the sides of such hills; as in Fig: 4.th [39]

VII.   My last Observations, are *Negative* (Viz.^{t}) First, that we can never finde, in any Hill whatsoevever, the Beds or stratas, to be equally parallel to *both*, or *all* the sides thereof; as described in Fig: 5.th but alwayes Scituate, as is before described in Fig: 1: 2: 3: & 4. and not otherwise.[40] 2dly That the hills and Mountains of the Earth could not be formed as the matter *precipitated* as is commonly Supposed; Because if soe, the hills would be allwayes found posited as in Fig: 5.th[41] or else confusedly without order, as in Fig: 7.th[42] In neither of which Positions we never finde them; and therefore they were not so formed and raised; but were certainly thrust up, by some internall force, after the Matter was hardned as aforesaid. And seeing these things are the Basis, on which I build my Hypothesis; 'tis not amiss to add a word or two for their further confirmation.

First then That the Matter was *soft* when these Shells were immassed therein, is manifest, even from the very walls of our houses; for we all know that the hair and straw, that's found therein, must be mingled therewith, when it was soft and tender, and not afterward: soe in like manner, the mettle of which any Molten Stattue is made, must be *thin*, and *fluid*; when cast into the Mould: And the Wax alsoe, must be melted; before it can receive the Impression. All which being soe plain and evident, I need do no more, but leave it to the Reader to make the application.

Secondly, That the Earth was *Smooth* and *even*, and no Hills raised when these shells were incorporated therein, or mixed therewith; will be likewise evident, From what is Visible to common Experience (Viz^{t}) Because the hills and Mountains in which they are found, *are now* much higher than the Ocean, wherein such fish are Now bred; and being soe posited, 'tis naturally impossible, that the Water and shells should come up thither, for the latter to be [p. 17] immassed therein:

And therefore the Hills at that time, must of necessity, be noe higher than the Waters then were, or now are; as was before proposed. But to be more particular; we have an Instance in the Ile of *Portland*;[43] where, near the Surface of the Earth, and to the depth of 60. or 100. Feet, may be seen Thousands of stones, (some of 40. or 50. Tunns a peece) soe full of shells throughout the whole Mass, and some soe full of their

Impressions (only) as that it seems difficult to judge, how more shells could be mingled therewith, or Impressions formed therein; As appears by Fig. 2^d. from one of which, I delineated the Rock therein described.

Now whether from these considerations, it be not more Reasonable to conclude, That the matter of this Island (and soe of all the rest of the Hills and Mountains of the Earth) was, at that Time, (to wit) when the Shells were mingled therein noe higher than the Bottom, or at most, than the Surface, of ye Water; Than it is to conclude that the shells, and water, were soe unnaturally raised 420. Foot (for soe much I found its perpendicular height) higher, than the greatest Tide or Flood, dos ever rise up to: You know the story, That the dirty Raffter in the Roof of the Barn, might *formerly* be down with the Cow, tho' the Cow, could not *then* gett up, to dung upon the Raffter: soe the Matter of the Mountains, could, and must of necessity, when soft, be even or Level with the Bottome, or at least with the Surface of the Water; tho' the Waters were never so high as the Mountains now are, since they were hardned, and raised; unless at the *Deluge*,[44] and then the Rocks were too hard to admitt them thereinto, as Will be fully evidenced, when we come to spake thereof.

And being describing the *Position* of the Bedds or Ranges of the Earth, there is one thing in these Observations that I have frequently thought might be of publique use, in the Digging of Quarries; or mining for coals, or Lead, &c.[45] And it is from Fig. ye 3^d. In which if you suppose the Top of the Hill at A. to be 300. foot high And supposing at **G**. you should digg 130. foot deep (to wit) to 2.B. [p. 18] and there finde a Mine or Quarry. Now you see by ye said Figure That if you open the Earth but 3. or 4. foot, at **B**.1. (on the side A.B.C.) you will then be in the same Debth with respect to the side of the Hill **A.G.A.** as you were, when 130. foot under ground, at **G**.2.**B**. And very probably you will there (to wit) at **B**.1. meet with the *same* Quarry or Mine, as at **G**. when at the debth aforesaid. And soe in like manner, that which is Level with the Horizone, as at **C**. and A. is nevertheless in respect of the side **AGA** above 300. foot Deep at **C**. tho it be there, but the very surface of ye Hill, and Levell with the **H**orizon as aforesaid.

Haveing thus described the figures, and also drawn and confirmed the Inferences, I shall Add these few naturall Propositions; as Postulata's, to the following Discourse.

Ist All Matter being without life, or not Modified or imbodyed; is now, and at all Times, in its *Caotick* Estate only; and is to be soe deemed and taken, untill it is assumeing, or hath assumed, a Life, either *Animall*, *Vegetable*, or *Minerall*.

II^{ly} As no Animall or Vegetable, can naturally be generated, but from *Caotick* matter only; so neither can they be Sustained; but by and from the same.

III.   As all *Matter*, by Generation & vivification, dos cease to be Caotick; soe all Bodyes, by Mortification and putrefaction, are thereunto again returned.

IIII.   As there can be no Life continued, without a *Regular* Motion; soe there can be no Motion, without more or less degrees of Heat; and as heat and motion, are

Inseperable; soe is that which is produced by y.ᵉ violence of Excess of Motion, (Namely) Fire, and light.

V. As no *Gross* or Fixed matter, (only) can admitt of Intrinsick Motion; soe neither can any *Thin* or *Fluid* matter, (only) admitt of any regulated Motion; Therefore all animated Matter; (or Bodyes) doe and must (besides the Active part) consist of *fluid* and *fixed* parts; the first to admitt of Motion; and the latter to regulate y.ᵉ same.

[p. 19] **VI.** As the Naturall Motion, that is *continued* in a liveing Body, is the same in kinde, with that which give^it i'ts *first* forme or Modyfication: soe that which gave it i'ts first forme must needs be of the same kinde, with that which is *continued* therein, for its Subsist:^ance or Sustenta:^tion v

VII. As all things in the *Chaos* (as shall be afterwards manifested) were thin and fluid, soe being imbodyed, that universall body, must consist of some parts more dense, and others more fluid, or rarified; And as the coagulated and uncoagulated, doe make the *Lactea Materia*; soe the Dense and fluid, or the Celestiall and Terrestriall orbs, and transparent Æther, doe make or constitute, y.ᵉ whole universe, equall to the *Chaos*, Nothing can be added *to*, or taken *from* the prima Materia, for it can only be converted or discriminated, into matter or Bodys of different Denominations.

VIII. As all *Putrefaction* is but a Languid and gentle Fire, so all Fire is but a more swift and violent Putrefaction; y.ᵉ end of both being to seperate y.ᵉ three principles of Matter, in order to fitt them again for Generation.

IX. As the Cloudes, or any other Chaotick Matter, cannot imbody itself into Vegetables or Animalls, whilst it is *moved* by y.ᵉ Air &c. but must be at rest before it can putrifie, and become of a Viscous Nature, soe neither could y.ᵉ earth and plannetts &c, vivificate, untill they had taken up their Centre of Rest; and become Viscous and Mucilaginous likewise.——

X. To Sume up all in Brieff; All matter in its Forming is divided into these three grand, or Universall Principles (viz:) *Mercury*, *Sulphur*, and *Salt*, or Active, Passive, and Fixed, or Musculous Matter;[46] of which the *Active* part is allwayes Inclosed within the Interiour Part of the Body Animated; and doth Naturally consist of an Intrinsick self=moveing, varifying and condensing, or dilateing and contracting Nature; by which, it is capacitated, to act upon the fluid or passive part thereof; soe that the s.ᵈ active part can be no less than the life, Spirrit, or soul of that body wherin it is inclosed as is [p. 20] evident by the Motion of the heart whence life, or the circulation of our Blood, is produced. I might add many more of this kinde but designing Brevity shall proceed.

Haveing thus premised and intimated, from what Terrestriall *phænomena's*, and philosophicall *Postulatas*, the ensueing discourse, is drawn and Established; I now come to consider, what that Materiall substance was, whereof the Heavens, or at least the Earth, were made; and of what it did Originally consist; according to the words of *Moses*, and our observations in naturall things to this Day.

By the account then which *Moses* has given us of the *Chaos*, or first matter, it plainly appears, that it consisted only of *Water*,; for he tells us, "That the spirrit of God "moved upon the Face of the waters; and God divided the Waters, from the Waters; "and that he separated the waters, that were under the Firmament; from the Waters, "that were above the Firmament; &c, Gene: Chap: 1.[47] And that this has also been the "Opinion of divers Philosophers, is also evident: In particular, from the words of *Basil* "*Valentine*, in his last will and Testament page: 95.[48] in which he tells us, That God hath "made the first seperation, according to his Word; "The Spirrit of y̌ Lord, moved "upon the *Waters*; The whole Elementall Body, hath been *Water*; But the spirit of the "Lord *Zeboath*, hath divided it, and fashoned the Earth, from the Mudyness of the "*Water*; And therein all Metalline Fruits, that ever were created, and generated under "ground. All These were first *Watter*, and may be reduced again to Water: All other "Creatures, be they Animalls, Vegetables or Minerall, are produced from y̌ first Water "&c.

Seeing then it is evident as well from these testimonies, as from the Pænomena's now Visible in the Earth, That the Matter thereof (as has & will be manifested) was at first soft and Fluid: And since *Experience* confirms what is asserted, (Viz̟t.) That all things may, by a proper Calcination, be converted again into Fluidity; We may therefore reasonably conclude, That the first Created Matter or Substance whereof the Heavens and the Earth, and all things therein Contained, were made, Was not from seeds, or Solid Attoms of different weights &c. [p. 21] But from a *Similar, Rare, and insiped Liquor, or Water*; consisting of a Vast and incomprehensible Orb, as to its form and Quantity; and of a Crude Lifeless or indigested Nature, in respect of its Quallity. And as y̌ simple Water of a Cloud, dos to this day, potentially Contain the Bones, flesh, Forme, Substance, spirrits, and senses; of all *Animalls*, and *Vegetables*, thence produced; soe this first Chaotick Liquor, did potentially containe, all the *Cælestiall* and *Terrestriall* orbs, and other Created Bodies, of what nature or kinde soever; which matter being by *God* the maker thereof, imprincipled with an indispensible Nature, Faculty, or Power, of imbodying or Animateing it self; it did by virtue thereof; *first* Engender and begett therein a Germinateing Heat, whereupon *Motion* ensued; and from thence proceeded a Secretion or discrimination of the *prima Materia*. And to Illustrate our meaning herein; This Seperation, (as it relates to the whole Chaos or Universe,) may be resembled to *Milk* actuated by any acid Quallity; which will thereupon, (by ascending & descending,) soe divide or Seperate it self, as that one part will become more *Opake*, and *Gross*; and y̌ other more *rare*, and *transparent*; as by the **VII**.th Postulatum is more fully declared; To which Job, in Chap: 10.V. 10[49] rightly alludes when he spakes of the Generateing of that Animall which is y̌ Epitome of the whole Creation.

And as this Germinateing Heat, attained greater Power, and Energie; according to the Process, of Naturall Heat, and Motion; it formed and disposed those discriminated

Parts of matter, whilst fluid, into such Globular shapes, as doe spontaniously result from all Suspended Liquids; (as may be evidenced by throwing a Bason of Water into the open Air; from whence many hundred Globular Sphers will be instantly produced,) and from thence Namely, from y.ᶜˢᵈ propensity in all fluid matter, the Bodyes of the Earth, Sun, Moon, and Numberless Host of Heaven, Attained their Naturall, and most selfsubsisting Formes of *Roundness*; of which we see them Universally consisting unto this day: Being also disposed, and suspended in [p. 22] that Transparent, unmoveable, Cristalline, and imperceptible *Æther*, or remaining Water; which fills or posesseth, all that Vast Roome or Space, which the said first matter occupied or possessed, before y.ᶜ separation was made; as may be inferred from the former Similitude; for the Lacteall Matter, is the same in its Dimensions, *after* y.ᶜ seperation, or Discrimination of its Parts, as when promiscuously mingled together; as is also asserted in Propos: **VII.** beforegoeing.

But for asmuch as those *Cælestiall* Bodyes are too remote, for our Observation to Anatomise; and seeing y.ᶜ Process of Nature, in Generating y.ᶜ smallest Animall, is the same, or applicable, to that of the Greatest; I shall therefore proceed to what I chiefly intended, being more particularly to treat of this Globe of Earth whereon we Live. And supposing, (as we reasonably may) That at its first Forming out of the Grand Chaos; It did then consist of a round, smooth, opacous, Liquid, Forme & subsistance, (see Prov: 8. V. 24. 25. &c.)[50] I shall in the following lines attempt to shew, how it came to be generated, & converted into that Diversity of matter; I mean of stones, Rocks, Water, Cley, &c. into which it is now, so visibly Changed. And for the Foundation of my Discourse shall lay down this proposition (vis.ᵗ).

That the Matter of which all things were made, being by the Author thereof impregnated (as aforesaid,) with a Vivifiyng or selfformeing Quallity; That Part thereof which was assigned, to constitute this Terrestriall Globe *whereon we live, was* generated *and converted into a Liveing Body; conformable, in respect of its* Vitality *unto that of a Liveing* Animal: *And that* y.ᶜ *Converting the said fluid matter;* into Rocks, Stones, &c. *was gradually carryed on, & accomplished, as the Body attained, its maturity and perfection.*

Now for the Illustrating and Explaining this Proposition I have transcribed the Generating of an Animall from *Sʳ: Kenelm Digby's* Treatise of bodyes, Page. 275;[51] that the same may be applyed to the Generating the Earth.

"To Satisfie our selves in the *Generation* of *Animals*, it were well (says he) if we "made our remarks in some Creatures that might be continually in our power, to "observe in them, the [p. 23] course of Nature every day &c. Which may be done by "meanes of a Furnace, so made as to imitate the warmeth of a sitting Henn."*. In

---

*. *I made my Observations more easie (*viz.ᵗ*] By putting 8. or 10. Eggs under such a henn, and takeing thence one at a time.*

"which you may lay severall Eggs to Hatch; and by Breaking them at severall Ages,
"you may distinctly observe, every hourly Mutation in them: The first will be that on
"one side you will finde a great resplendent clearness in the white; after a while a Little
"spot of red matter like Blood will appear, in the middle of that clearness fastned to the
"*Yelk*, which will have a Motion of *opening* and *shuting*, soe as sometimes you will see
"it, and Straight again it will vanish from your sight: And indeed at first it is so little
"that you cannot see it but by the Motion of it; For at every Pulse as it openeneth [sic]
"you may see it; and immediately again it shutts in such sort as is not to be
"discerned—From this red Speck after a while, there will Stream out a Number of
"little almost imperceptable red Veins; at the end of some of which in time there will
"be gathered together a Knott of matter, which by little and little, will take the forme
"of an Head; and you will ere long discern Eyes and a Beak in it: All this while the red
"spott of Blood grows bigger and Solider, 'till at length it becomes a flessly Substance,
"and by its Figure may be easily-discerned to be the heart, which as yet hath no other
"Inclosure, but the substance of the Egg, But by little and little, the rest of the Body of
"an Animal is formed out of these red Veins, which stream out all about from the
"Heart; And in process of time, That Body incloses the heart within it, by the Chest
"which grows over on both Sides, and in the End meets and Closes its self fast
"together. After which this little Creature soon fills the shell, by converting into
"severall parts of its self, all the substance of the Egg, and then Growing weary of so
"Streight an Habitation, it breaks Prison and comes out a perfect formed Chick.

   "To this may be added an account to the same Purpose, from the Works of the
"Hon:^rable *Rob: Boyle* Epitomized, page 73 (vizt):^52 "That the Rudiments of the Chick,
"Lodged in the *Cicutricula*, are nourished only by the White, till it becomes a great
"Chick; the Yelk being reserved, as a stronger Nourishment, till the white is spent; and
"then the Chick is able to digest it, and in Effect the Chick seems to be furnished with
"Head, Wings, Beak, and Claws, before the yelk is touched. Lastly (says he,) it is not a
"Little to be admired, that so *soft* and *similar* a Liqour as that of an Egg, should be in so
"short a time, changed into a Chick, endued with Organicall Parts, of different
"Fabricks, and, Similar ones, different in Feæture very much one from the other &c.

   Now that it may not be objected that these Changes and Pulsations, are wrought
in the matter of the Egg, by means of this artificiall way of hatching, or by any Virtue
derived from the hen, when therein Engardged; any otherwise than only by the
applycation of heat thereunto; It is to be noted that the same Operations, are wrought
in the Eggs of an *Ostrich*; which being by them laid in the Sand and warmed only by
the *Heat* of the sun, is as naturally hatched, as any Eggs under any Fowle whatsoever. I
shall therefore in the ensueing Discourse, mention the Ostrich instead thereof, and shall
conclude this head, with a short Relation of what, I have my self further Observed
(vizt),, *That in the spring and summer seasons, near ye sea shore, there is frequently found, a*
certain matter or substance called, in these parts by the name of Bulls; much like ye White of an

Egg in Taste and Colour; and in Bigness, about the Breadth of a mans hand, in Forme Circular, and thin on the Edges; but in the middle, on the uper side thereof, which is flatt, there appears as they float in y.ᵉ salt water, a Circle somewhat whiter than y.ᵉ said matter. This Circle I have seen to dilate and contract it self in a constant Motion; somewhat swifter than the Beating of our Pulse; But into what Body this matter is thereby Engendered, or Animated I know not; however it is not to be doubted, but some sea Animall or other, is thereby, Especially, at such a season produced.[53]

Haveing given you an Account, of the *Generating*, [p. 25] *Oviperous and other Animalls*, and indeed therein of the Generating all other Animated Bodyes whatsoever, for as certain as the least line BD. ffig: 6.ᵗʰ[54] is equall in Power, to the greatest Line A.C. (see Eu: li: 3. pro: 35.)[55] soe certain it is, that the *least* Animall, is Equall, in its self=moveing, self=preserveing and self=Multiplying facultyes, as the *greatest* Animall whatsoever, I say, haveing thus done we will now proceed to apply the Same to y.ᵉ Generating the heat and motion, within the Body of the Earth which we told you was vivified with a Life resembling that of an Animall; whose *vitallity* we all know, dos principally consist in y.ᵉ Motion or circulation of its fluid parts; as the Effects of its internall Pulsation.

Now inasmuch as y.ᵉ Earth, is one of those Bodyes, that were Generated without y.ᵉ applycated heat, or Assistance of any other Naturall Body; It will therefore follow, that she must by the infused plastick Power before mentioned, accomodate herself, with such an intrinsick Heat; or otherwise it could not *Naturally* (for we are only so speaking) be Quickned and Vivified, as is before proposed And that such liquid, indijested, or unanimated Mater, will gradually Engender and produce, such an intrinsick Heat and Motion, is Evident from all Mucculent or Slimy Filth, whence wormes and Vermins are to this day, frequency Engendred; as also from Worts or Beer (though cold) put together in a Guile=Tunn; and moreover in Stacks, or Reeks of Hey, which are thereby, Sometimes, burnt to Ashes.[56]

Now this gentle Heat, and therewith Motion, being (Like the said Moving speck in the Egg) Engendred in the interiour Part of this liquid Globe; it had the same Effect thereon which is common in the Works of *Generation* to this day; (viz.ᵗ) to Discriminate, and Conjoyne; the three then promiscuous parts of matter, to themselves respectively; for although this Orb, was at first (as is said) somewhat like the Water of a Cloud, which to Appearances, seems of one homogeneall Nature; yet after it had lain Brooding or compressed together, and [p. 26] thereby Engendred the said Heat, and Motion, it did there=withall, not only Viscate, and Coagulate; but also Segregate Discriminate, and Dispose; the *Active*, *Passive*, and *fixed Parts*; that are potentially inherent in all such Matter, (according to the **V**.ᵗʰ and **X**.ᵗʰ Aphorisme before mentioned) in manner following; (that is to say,).

Those that were coagulated and Converted into **Fixed** particles (i.e.) stone &c. were disposed, and transposed, in such manner as to Give Shape and forme, to the

Exteriour and interiour parts of the Body of the Earth; (being then in its Generating or Animateing Estate).

Those that were by Nature, designed for y.ᵉ *Active* part; were collected and transposed in such sort as to Constitute the Cordious part, or (if it might be soe called) the **Spirit** or Soul thereof. And at y.ᵉ same time, (for the universall or constant process of Nature, is to carry on the forming of all the parts of an Animall together,)

Those that were designed for the *Fluid* or *passive* part; were so disposed of, as that y.ᵉ Active part might give a regular Motion thereunto; Which being fully Effected the Body was compleatly Animated. And this will further appear by duly considering the aforesaid Example of Hatching the *Ostrige*; For as the Naturall Heat of the *sun*, Begetts an internall Motion in the Glutenuous matter of the Egg, which formes the heart, and therewithall the whole Body, with a *Cavity* wherein the heart may Dilate and contract it self: soe in like manner, the Motion that was Generated in the said Orb, did convert, transpose and conjoyne, the said conglaminated matter in such sort, as to forme this Independent Body of the Earth;* with a sphericall Cavity whereby the Active or Cordious part, might be capacilated to Pulsate; [p. 27] and retexerate the then whole *fluid Mass*, and thereby Vivificate or animate the whole Body; as before asserted, And truely were there not such a Cavity gradually formed; there could be no Motion, and Consequently no Life. For if the Body was (according to the 5.ᵗʰ Postulata,) All Solid and compact, there could be no *Dilation* and *Contraction*; which are the only Symptoms of Life, in all animated Bodyes whatsoever.

But to be more particular as y.ᵉ said viscous Matter of the egg, is converted into *Bones Flesh Sinues* &c: And the yelk into the *Bloud*; soe in like manner, part of the liquid Matter of the Earth was as is said, converted into a viscous Substance, and afterwards turned into *Cley*, which in process of Time, was hardned into *Rocks, stones, Flints, sands,* &c: which constitutes the Fixed or exteriour Part, or Forme of y.ᵉ Earth; as is intimated, and Sett forth in the Foregoeing Figures and their Descriptions; One other part thereof, was congregated or converted into a *Cordious* Active matter, and was disposed of, as aforesaid. And the third part thereof was converted into *Passive* Matter; being that which to this day constitutes the salt fluid *Ocean*: And that the Ocean is only passive, and not selfmoveing, is evident; in that, if any Lake or other Quantity, be seperated therefrom, it has no more power to move it self, than any Dead or Lifeless matter whatsoever; as is evident by the *Caspian* sea, and other Lakes. And now we are speaking concerning the disposing of the different matter of the Earth: It is to be noted, that it was not according to the strict Rules of Specifick Gravity,[57] as many in times past, and of late, would have it to be; For if soe y.ᵉ Minerall and most ponderous

---

* I call that an Independent Body which is generated of an Orbicular Forme, and is not radicated in any other Body, Nor doth depend upon any food for its subsistance; y.ᵉ contrary may be called dependent Bodyes.

Matter, would have been next unto, or used for; the Seiling, of the Sphericall Cavity in the Earth; and consequently, would have been soe deep in it, as that we should never have known what Gold, Silver, or Brass had been. But for=asmuch as dayly experience shews; that Bedds of Chalk, are under Bedds of stone; and Bedds of Chalk, Cley, Sand, and [p. 28] stones; are under Bedds or Mines of Lead, Tin, silver &c.: I hope there is no Roome for such Notions, as are so visibly confuted. And therefore we may reasonably conclude that the said matter was disposed and conjoyned according to the Course and Process of Nature, used in generateing the *Ostrige*, or in the forming of any other Animall; as before porposed. But perhaps it may be objected; That if the Earth, be an *animated* Body; why was it not converted into *Veins*, *Leggs*, *Bones*, *Members* &c: as well as other Animalls.[58] To Which,

I answer, the Author of all Things, haveing made some creatures to be independent, or not needing the Assistance of other matter for their preservation, has therefore made such; of a perfect *orbicular* forme; and as it were Hermitically sealed up their Spirrits, or active parts, within themselves, for their perpetuall Preservation: But haveing on the contrary made other Creatures, soe porous; as that their Heat and Fluidity, are perpetually avapourating or flying away from their Bodys; (which renders them always needing a supply, for their preservation,) has therefore ordained, that these Avapourations, Decayes and Volitions, shall be supplyed and Compensated, by the Heat and Nourishment, a rising from the *putrefaction* of other liveing Bodyes; (or such like Matter,) in its being converted or turned into a Caotick Estate; and by the re= incorporateing or re=animateing, part of y:e same Matter again, into the very Essence or Existance of that Body whereinto, and whereby, such matter is Received, and Concocted. And this appears by his haveing formed Various sorts of Creatures, and furnishing them, respectively, with members, Caveties, and Capacities; for obtaining, receiving, concocting, and converting, such Bodyes, or matter, to, and for, their own peculiar nourishment, and Procreation. And hence it is, that all Terrene Animalls, and Vegetables; (being dependent Creatures) As they had their Generation, Life, and [p. 29] Being, *from* the Earth; soe they must also unavoidably have their food and nourishment, from y:e same likewise; but as these parts are thus necessary for the Existance, and support of those Bodyes; soe they are altogether as superfluous and needless, for such as are independent; as the *Earth*, *sun*, *Moon*, and *Starrs* are: And seeing he that is infinite in wisdome and Power, has made Nothing in vain, It cannot rationally be supposed, That he should make any thing, either to Long, or to short, deficient or superfluous; and therefore it would be absuridity in us to imagine, he should have formed the Earth with *Veins*, *Bowells*, Leggs &c. Because they are altogether needless in respect of its being, or future Sustentation: For if we should imagine her to have *Leggs*, Whereon should she tread? Seeing *the Lord hath Stretched out the North over the empty places, and hanged the Earth upon Noth*ing Job. 26. 7.[59] If a *Mouth*, whence should she receive food or whereinto should she eject it: and soe of the rest;

The consideration of which independancy, will lead us to see the naturall Cause, Why the Earth, and the other Orbs have that perfect Forme of Roundness; which they do so visibly retaine: But the *Cause* why the forme of all other Animalls, are soe different from y.ᵗ of the Earth; will more fully appear, if we compare the manner of the *Subsist*ance of one Creature, with another; which may be done in this wise:

No common Animall can Subsist or receive nourishment from the Earth in *one* place only; As a Vegetable by the fall of Raine, naturally dos. But must remove themselves, to seek their Food in divers places; and therefore they are Variously furnished with parts, Members, and abilities; as their wants & necessityes, do Severally requir. And hence it is, That a *Worme* tho' in respect of its internall Heat and Motion, or Vitality is as has been said the same in proportion with the great *Leviathan*,[60] or any other Animall; Yet the Roots of Vegetables, or the humidity of the Earth, being assigned for its food; and she also placed therein, [p. 30] is therefore formed without Leggs, or Whings, as being needless for obteyning its Subsistance. But oviparous and other Animalls, being such as are unable to subsist, in one place, or on any one thing; are therefore accomodated, not only with Leggs to Walk, but respectively with Wings to fly; as well for self=subsistance, as for self=preservation: soe in like manner, *Doggs*, *Woolves*, *Lyons*, *Hawks*, and such other Creatures, as are appointed to Live by Preying on other Animalls; have their Agillity, strenght, and Swiftness of Motion, proportionably assigned them. And as we thus finde, the Causes or Reasons of the Difference, in the Formes, and members of Creatures in generall, are only Such as theire Self=Subsistence, and Self=preservation doe necessarily require; Soe in like Manner we may find, That the Severall *Senses* (and the Degrees thereof, where with they are Soe variously furnished, is only on the Same consideration likewise; For if the Beasts of prey, had not *Sight*, and *Smell*, added to their Agillity, they could not obtaine Food For their Subsistance; nor distinguish what was fitt to be Eaten or what to be refused; From the want whereof, theire Species, would unavoidably cease, and be discontinued; and soe of the rest of the Senses. Nay this is also manifest, even from Man himself; who in Respect of the Forme of his Body, (he walking Erect, and his Head, tho' not his Thoughts,) raised So far from the Earth, is thereby the most unfitt of all Land Creatures to feed on Vegetables; and by the paucity of his Feet, the same, in respect of his Obteyning Animalls; and therefore to supply these Defects, he's soe admireably furnished not only with senses, common to other Creatures, but also, with crafty Wisdome and Knowledge; insomuch, that they are therein soe inferiour to him, as to become a prey, not only to his hungry Appetite; but also, many Times, to his *Shamefull* Excess and Ryott.[61]

[p. 31] And thus might we finde, the Naturall Cause of the *Forme*, *Members*, *senses*, *Wisdome*, *Ignorance* &c: of one Creature, compared with another; and also, that if they had (like y.ᵉ Earth) been formed perfectly round; they would not have needed or required any such Members, or senses for their Subsistance. But the Reason why they were not *soe* formed is because the liquid matter that produced them, (by its being posited in divers Lakes and Ponds upon the Earth, And not Suspended (as y.^e Orbs are,) in the open *Æther*,) could not be Globular; and Therefore Nature generated them, with diversity of Shapes, and Magnitudes; according to the Nature and Quantity of the Menstruum, whereof they were Engendred or produced: As also with Self=moving powers; because *Locall* Motion is indispensibly necessary, for all Animalls, that are not orbicularly formed; as has already and will in Sect. **IV**. be more fully confirmed.

But though we have Shewn that *such* Members, Senses, and Faculties, are Needless, in the Body of the Earth: Yet the *Heat & Motion*, or Pulsation; as also the *Cavity* found in all Animalls; are so essentially necessary to the Subsistance of the Earth likewise; as That it cannot be a Living animated Body without them; We will therefore in the next place shew, that these Qualifications and Capacities, are to be found therein. But before we enter thereupon, It may not be amiss to Acqaint the Reader; That altho' we have frequently proposed, the Earth to be an Animated Body; Yet our principal design, was only to finde out ye manner how, and when, the Shells, and other Marine productions, came to be immassed and mingled in the Rocks and Mountains thereof; As also what is the Originall or Naturall Cause, of the Flowing of the *Tides*: so that as, what we have intimated, in respect of the Earths being an *Animall*, is cheifly, to illustrate our conceptions, in Relation to the Said Enquiry: soe the Earths appearing to be Such, [p. 32] is but only as a Corollary, arising from the same Enquirie; and therefore what is otherwise Asserted, in relation to the premisses; may be granted as *true*, Whether the Earths being an Animall, be believed or rejected. But however this I must Assure You; That although I had Spent Severall years, in Contemplating and Reasoning, upon the said Phænomena's; Yet untill I had, (at a certain time, and place well remembred) clearly discovered the Body of the Earth, to be an *Animall*; I always found it impossible to conceive, how the Said Shells, Could, (without a Miracle,) be posited in the Rocks, and the Waters made to *Ebb* and *flow*; Though Now, not only the before mentioned *Phænomena's*, but all others, relating to the Earth, are, by supposing it to be an Animall, made plain and evident, and very easie to be accounted for: Seeing therefore, it dos so naturally, and unavoidably arise *from*, and depend *upon*, the said Enquiry; I should not speak the thing that is *right*, concerning the Work of *Creation* (Job: 42. V. 7.)[62] should I avoid calling it, what it so plainly appears to be; haveing then already shewn, that there is a perfect resemblance, between the Earth, and its Inhabitants, in respects of their Vitality, and Essentiall parts; I shall further Shew, that there is the *Same*, in respect of their Natures, and Productive faculties.

And in the first place, it is to be noted, That as the Heat, and Serous Sweat of the Earth; did naturally produce *Horses, Sheep, Swine, Doggs*, and other Creatures: So the like Sweat of Those Creatures do Still produce, *Fleas, Tiks, Lice*, and other like Animalls; . . . and Some of them (like the Earth), even of more than one Speices: And by the Same Analogie, no Doubt, but many of these lesser creatures do in like manner, produce and nourish others [p. 33] that are, to us, unknown and imperceptible.

Secondly, As the *Earth*, by its Sweat, (or what is thence produced), dos feed all Creatures, that did, or do proceed from it: So in like manner, the Said Animalls, do naturally feed and Nourish, all those that are produc'd by *them*, And as all those lesser Creatures, that are externally produced by *the* Sweat of common Annimals are of different *Species*, from those by whome they are produced: So those *greater* Animalls, that were Externally produced by the Earth; were, and are, of a different *Speices* from the Earth: But those Creatures, that are now *internally* generated; by, and from, Such a *Menstruum*, as that by which the Earth at first produc'd the said greater Animalls; (Namely) By, and from, the same matter, that is by Nature Concocted and Assimulated, to *that* Body, By which they are to be produced; I say, These are always, (Naturally) brought forth, of the same species with the **Producer**; and not otherwise. And therefore, as we have already Shewn, Why all Animalls were not formed Orbicular; so we may from hence further learn; That the Reason Why, they were not of the Same forme and *Species* with the Earth; was because they were Externally brought forth; and, if not wholy, yet perhaps, cheifly, by its own Serous Sweat only; and not by the Same numericall Matter of which its own Body was produced. And therefore it could not produce them, as is said, but of a *different* Species; altho' the producer, and Produced, were both of them, perfect Animalls: And this is Confirmed, from the Production of such Creatures to this day, (Viz.ᵗ) In that a *Flea*, is an Animall; tho' it be not a *Dogg*: And a *Dogg* is an Animall; tho' it be not a *Flea*; and the same may be understood of the Earth: For as the Earth's Inhabitants, are Animalls, tho' not shap'd, as *it* is: so the Earth, is an Animall, tho' not shap'd as *they* are. [p. 34] By which it is evident, That com̄on Animalls and the Earth are both alike, or Assimulated to each other; as before porposed.

And as they are thus Assimulated in respect of the productions Upon their Bodies; so they are likewise in those that are Radicated within their Superficies: For as the Earth produceth Vegetables and Mineralls, (to wit) *Trees, Grass, Coals, Lead, Tin, Sulphur*,[63] and other living and Growing Excrescences: So the other Animalls, do in like manner produce, *Hair, Nailes, Feathers, Hooffs, Teeth*, and other Excrescencies, Radicated in and upon their Body'es likewise; and as these Vegetables, and Mineralls, are perpetuated, the first of them, by their Seeds, to supply the Species of those that will perish through Cold or otherwise And the latter, (to wit) the Mineralls, by their Roots only; so the Hair, Nailes, Teeth, Hooffs, &c. of these lesser Animalls are perpetuated, (Not only during the Lives of the said Animalls, but even after Death,) only by their Roots, likewise. Because seeds, are altogether needless, and useless, to such Excresciences, as have *no* Menstruum provided for them, as the Vegetables and Mineralls have; nor being exposed to Cold; as aforesaid. From all which it is evident, That the Earth, and its Inhabitants, are Assimulated to each other, as well in respect of their Natures, and Productions; as in their Vitality, and Essentiall parts. I shall now proceed to speak of The Earths Internall Heat and **Pulsation** before proposed.

Sect: II^d.

That Heat, Motion, and Pulsation are in the Body
of the Earth; as well as in other Animalls. Divers
Arguments from Nature, Reason, and Experiments proving
That the Moon hath not the least Influence over the
waters: Nor any Planet (besides the sun) over any
Terrestriall Animalls: As by a Register of the Tides,
Kept by the Author, for above two years Successively;
By which it appears, that they Flow in a Rogarithmeticall [sic]
Motion, and not according to the Moon's: That the Flowing
of the Ocean, is ye Cause of the Earths Rotation; and
that That, is the Cause of the moons Revolution;
illustrated by an Experiment. That the Moon has no
attractive power over the waters. And that the
Gravitating power, lately ascribed to her, when in her
Perigæon, is but an imaginary Supposition.

IT appearing, by what has been Said, That the Earth cannot be an animated Body,
without an internall *Heat and motion*, or pulsation; we Shall therefore endeavour to
prove that — — those Properties or Quallifications, are to be found in the Earth, as
certainly as in any other Animall; and the First of these to wit its internall Heat will
appear *First*, From the hardned Cley and Rocks, that are Scituate at the Bottome of the
Sea, and many hundred Fathoms deep in the Earth, where the Suns Heat, nor any
other, but that of the Eath, can be pretended, to harden the Same. 2^d: From the
Sulpherous Heat, frequently occuring in Mines, and other Subterraneous Depths; as
also from hot Baths, and the manny Sudden and constant Eruptions of Fire, which are
mett with, in manny Parts of the Globe. But this Heat in the Earth being [p. 36]
generally granted I shall not Spend time for Its Confirmation; butt proceed to what I
cheifly intend; being, in the *Second* place, To Shew, That the Earth, hath an internall
Motion or Pulsation; And though this is evident, from the **IIII**^th *Postulatum* or Aphorism,
(vizt) because heat and motion, are inseparable; yet I shall attempt to prove it more
fully, from the visible Effects it has in moveing the Waters in the great Ocean; and
Shall, in order there unto, Lay down this Proposition; namely,

"That the Ebbing and Flowing of the Ocean, called the Tides, *is caused by the internall
motion or Pulsation, always agitating within the Body of the Earth, or under the Ocean; And that
it is, with Respect to the Earth's magnitude, in a naturall proportion, to the* Pulsing, *in the Bodys
of all other Animalls; Which motion, as it is observed, in lesser, more swift; and in greater, more*

slow; soe in this *vast Body of Earth, it pulseth only once* in 15: Days: *being (as the effect thereof approacheth on the Respective Coasts) then commonly called Spring=Tides: All other* Ebbings *and* Flowings *therein, being but only from the the* [sic] *Surge, or consequents, of the said fifteenth Days* Pulsation. *And this I propose to prove, both* Negatively *and* positively and,

 First, **Negatively,** *That the* **Moon,** *is not the Cause, of the* **Ebbing** *and* **Flowing,** *of the* **Tides.**

Forasmuch as my chieff designe, in this Negative parts; is intended to unhing, an Antient, and almost universall Opinion; I cannot expect, but that the taske will be the same, with what is common to persons concerned in rebuilding antient Edeficies (to wit) prove more [p. 37] Labour to remove the *old* than to Lay the Foundation of the *new*; I must therefore begg leave to be more large on this head, than otherwise might be required. Now what this Opinion is, we cannot be ignorant; For we all know, Tis a common Notion *That because the Ebbing and Flowing of the Sea; (dos,* (as is pretended) *exactly corespond with the Motion of the* moon; *Therefore she is presumed, to be the only cause thereof.* Which I conceive to be more irrationall, than to ascribe the Motion of the *Moon*, to the Ebbing and flowing of the sea; and that, for these severall Reasons following.

First the *Moon*,[64] is by Astronomy found to be forty-five times less than the Earth; and but a Secondary Planet; and allowed to be of an Earthy Nature; and therefore destitute of any such ignious Rays as do usually proceed from the sun: And being but a *Sattellite* to the Earth, is therefore, (according to the Rule of the Major governing the Minor) more liable to be influenced by the Earth; if there was such a thing as Influence in any of the Orbs, (besides the sun,) than to be presumed the Lady Governess of the greatest Part of the Earths Superficies: And indeed of that part, which keeps it, from returning to its first *chaotick* and lifeless Existence. for Experience tells us, that if the Blood in our bodyes, be wholy stagnated; the Life is thereby fully extinguished; nay I would appeal to any rationall Person; whether it may not be more probably concluded that the vast Bodyes of *Saturn, Jupiter*, and the *Earth*, do give motion, to their respective Satellites, (of which the moon is the Earths') than that such small Bodyes, should Influence, or give Motion unto them, or any of such their essentiall Parts.

The second Argument is this, (Viz<u>t</u>.) That the *Being, End*, and Effect, of all internall motion, in all animated Bodys; is appointed and appropriated to, and for, y<u>e</u> proper use and Subsistence of the *same* Body only. Hence the flowing of the sea, the pulsing of our Blood, the igneous Vapours or Rays of the sun &c. are necessary Effects of their [p. 38] internall Motion; and are peculiar to the use, and Subsistance of these Respective bodyes only; for the *Sun*, dos not Send forth his burning rayes on purpose to Vegetate the fruits of the Earth; as some may imagine; But as it is an Effect of its own internall, and ambustuous Nature: It being the Same to the *Sun*, whether the Spring or fall, dos, or dos not, correspond with his presence or absence; or whether his Rais, do or do not touch the Earth; (for there are as it were infinitly more that do not come near it; than those that are darted therupon as appears by Fig: 10:)[65] Soe in like manner; tis the Same to the flowing of the Sea; and to the Motion of the moon; whether they doe; or doe not correstpond with each other: And here it is also to be Noted, That altho' the Sun has Power (by insinuating its Rays, or rather ignious Vapours into the Terrestriall Waters) to rarifie it, even almost in a parpendicular Elevation, as is common in Raising *Watter=Spouts*, and Vapours both at Sea and Land; Yet there is no power or Influence in the *Sun*, (nor yet Volgarly ascribed to him) whereby it can in the Least, incline the Sea to Ebb, or Flow towards any point of the Compass whatsoever. And if the *Sun* that has Such visible Influence, to Exhale the Watter, in Such wise; has no power to cause the Ebbing or Flowing of the Ocean; How the *Moon*, that has not power to Send the least Ray of Ignious Influence upon, or to draw the least Vapour from the Earth; not to keep it up by Night, when the Sun in the Day time has rarify'd it; I say, how the *Moon* should have power, to move so vast a Body of water, is to me a seeming impossibility, and very unreasonable to beleive.

And as these things are very improbable, if not impossible: soe is also that common received Opinion of the Moons *attracting* the Ocean as the Load=Stone dos Iron; or as others; That the Waters do naturally move towards the moon. For the weakness of both these Suppositions, will plainly appear, by compareing the constant, and regular Influence of the *Magnet*, over the Needle; with what is falsely ascribed to the Moon, over the Ocean:[66] In which 'tis manifest. I[st]. That that [p. 39] *LoadStone* which will lift up, but half a pound of Iron, will if joyned to halfe a Dozen pound, hung by a string, or posited in Equilibrio, (as the Ocean is) will, I say, draw it which way you please; either backwards or forwards, or retain it, where you fix or posite the stone, (within its influence) and not otherwise. But the motion of the Moon, if compared with the Tides, will be found far otherwise, (Viz.[t]) to be one half, like *attraction*; and the other half like *Rejection*; For 'tis well known, The very Tides of Flood, do, as constantly, twice in 25 hours, runn up the English Channell, (and else where) as Swiftly, [from] that point of the Compass which the moon is on; as they do alternatly [towards it] whereas if the moon had any Magnetick Virtue, it could not admitt of such manifest contrarietyes. And as there is not a Similar Conformity in their Motions, in *this* particular; soe it dos as evidently appear, by the Waters of y[e] *Caspian Sea*: (which in Quallity, is like the Ocean, in saltness; and in Quatity, above 2000. miles, in Compass,) For if she had such Influence; or if the Waters did naturally move or Incline, towards the Moon; they would in that vast Lake; either be carryed round, or move themselves, to every point of the Compass, according as the Moon was posited: But instead of Such a Conformity, or Influenceing, there is not the least Motion found therein; (other than what is accidently given it, by the Winds) although the Moon be at full, or Change, or what other Estate soever. And what is here said of the *Caspian sea*, would be also found in *Pontus Euxinus* and the whole *Mediteranean sea*; were it not for the small connections, made by the *Streights* Mouth and the *Helespont*. Now that these (comparatively) small seperated Quantities, should not be moved, nor Influenced by the Moon; and, at the same Time, the vast Ocean, allowed to be agitated by her; can be no otherwise granted; than that the same gust [p. 40] of Winde, which can drive, the largest Vessell, has not power to move the Smallest Feather: Or That that Magnet, which has power to lift up a pound of Iron, has not power to take up a diminutive Needle: The Absurdity of both which dayly Experience manifesteth. And as to the great and vulgar Argument, (Viz.t) *That the Moon must Govern the Tides, because they do exactly correspond with her Motion,* (tho' I hope by and by, to convince those who have taken this upon Trust, that there is no such exact Correspondence between them;) Yet in the mean Time; I answear, that if there were such a conformity, it dos not thence follow, that the moon is the *cause* thereof; and to Illustrate the matter, Lett it be supposed, that an Ingineer at *Bristoll*, should make a pair of Chimes; soe, as to goe every Time, it was full sea at London Bridge: Another at *London*, should make a Clock soe, as to strike every Time, it is full sea, at the *same* Bridge; surely it cannot follow, from this Correspondence, that either of these Movements, had any Influence, or were dependent upon each other; but that they were Severally agitated by their own *respective* Causes: And the same may be understood of the Motion of the Tides if compared with the Moon: From whence I presume it dos plainly appear, that the moon, has no *magnetick* virtue, or Influence over the Waters, notwithstanding the said conformity if allowed: And tho this may be Sufficient, to convince some persons, yet, that a full confutation, may be given to this generally received Error; I shall also shew, that she has no *Soler=like* (or other) Influence whatsoever, by which she can in the least affect this Globe of Earth, or any part thereof; and in the first place, shall lay down this self=evident *Postulatum* or Proposition, (viz.ᵗ)

"*That* no Plannet *or* Fixed Star, *nor any other* naturall *Body whatsoever, (devoyd of* "*sense) can* influence *or* affect *another like Body; unless such Bodys, or their Atmosphers,* "*do actually* [p. 41] *touch each other: or else both of them,* do touch *some proper* "Medium, *that can convey the* action, *or* Influence *of one, of them, to the other.*⁶⁷

Now that this Globe of *Earth*, not its Atmosphere, is not touched, by the Body of the Moon, or any other starr, nor by any of their Respective Atmospheres, (unless the suns';) and that the Sun's Atmosphere, is not a proper Medium, to convey any Influence, *to* the Earth, from any other Celestiall created Body; is what I purpose to demonstrate; and in order thereunto, that things may be plainly evidenced, I have, (as the Room would admitt,) Subjoyned a Scheme, in Fig: 10ᵗʰ of the *Position* of the Planets, and the proportion of their Magnitudes, compared to each other; according to that naturall, and well approved System, of *Copernicus*; By which it plainly appears,

That the Body of the Moon, tho' she be ten times nearer the Earth, than any other Planet; is nevertheless, allowed to be above 114000. Miles distant from it; And their Atmosphers but very littleless. For tho' that ot the *Sun*, is so vastly great as to extend from B. to A. in Fig: 10.ᵗʰ (viz.t) from its own Body, to the Body of *Saturn*, and perhaps much farther. Yet the Sublimations of the Earth, as has been, by *M.ʳ Hally*, ingenuously computed;[68] do not extend above 45. miles in height: And the Moons, as by *M.ʳ Huggens*, much less.[69] From whence it plainly appears, That y.ᵉ Body of the Earth, not its Atmosphere, cannot be touched by y.ᵉ Atmosphere of the Moon; nor by any other Planet, or their respective Atmospheres (unless the *Sun's*) and consequently they cannot, by reason of their remote bodily position, in the least affect, or Influence each other; as was before asserted.

I now come to prove, That the Sun's *Atmosphere*, is not a proper *Medium*, to convey that which is called the Influence of the Cœlestiall Orbs, to the Body of the Earth. But in the First place Lett us explain, What is to be understood, by what is Vulgarly called *Influence*,[70] and to *Which*, there has, in all Ages, been so Secret, and so Occult Vertue ignorantly and falsly ascribed: Now to guide us herein we have the *reall* and *Sensible* Influence, of one of the Planets; namely, the *Sun's*; By which, if we rightly understand, what his Influence is; we may, by comparing it, truly finde out, what is meant, by that ascribed to the rest of the Planets.

The *Influence* then of the Sun, is, (in breiff) nothing but the ignious Vapours or Sublimations, arising from that burning Globe of Fire; which perpetually ascend from and, (like our Watry and Airy Vapours,) precipitates again upon its own Body.

Now inasmuch as this reall and Sensible Influence of the *sun*, is nothing but the said Vapours, & Sublimations thence arising; and forasmuch as we have Watry Vapours, in like manner, perpetually ascending from, and returning upon our Earthly Globe, (which in imitation of the Sun's might be called the *Earths Influence*:) And Seeing the Moon and the rest of the Planets, are found by Observation, to resemble the Body of the Earth; as well in their Nature, as in their Forme. What may be justly concluded from this Analogies. But that by the **Influence** ascribed to the Planets, we ought, (in a Word), to understand nothing else, but the Watry, or Firy *Vapours*, arising from their respective Bodys: Since then it is evident, that those Vapours, or Sublimations, must be what is meant, and to be taken, for the **Influence** of the Planets; And Seeing those respective Vapours, cannot, as aforesaid, touch each other; we will examine, whether the Sun's Atmosphers, be a proper *Medium*, to convey them to the Body of the Earth, or to its Sublimations. and

In the first place, it is to be remembred, That we have in the Scheme annexed, (See Fig: 10.ᵗʰ) already Shewn; That the nearest of the Planets (namely) the *Moon*, is by Scituation, distant from the Earth above 114000. *miles*; [p. 43] what is further evident, is from their Position in the Line ♄. ♃ ♂ &c.[71] wherein it is to be noted, That if the Vapours, of any of the Plannetts, could be conveyed, by the Sun's Sublimations; yet those that belong to the Moon, (when in Opposition, to the Sun) as also to *Mars*, *Jupitir*, and *Saturn*, would, instead of being conveyed *to* the Earth, be carryed farther off from it; as the Ignious Vapours of the Sun, ascends towards the bounds of its Atmosphere, at A.A. &c. soe that there is only the Vapours of the *Moon*, *Venus*, and *Mercury*, (and those but only when they are in conjunction with the Sun) that can be pretended, to be convey'd to the Body of the Earth: For when the said Plannetts, are removed, but 4. or 5 Degrees. from the Conjunction; their Vapours, if conveyed, *in* the Said *Ætheriall* matter, would not touch, or come near the Body of the Earth; as is manifest by the said Diagram; where the Speck, or the planett *Venus*, is removed to the Letter D. in which position, its Vapours could not, in the least, be conveyed by the sun beams, to the Earth, to influence the same; and so of the ☾ when removed to F. as is before asserted.

But besides all this, 'tis very ridiculous to think that the Vapours of any of those Earthly Bodyes, if, even (when they were in a right line, as at **SE** in Fig: 10.ᵗʰ) should be convey'd or carryed, for soe many hundred thousand Miles, in such a different Sublimation, as the Sun's is; and after all, he brought back again, in the *same* Quantity, to their respective Bodys; (which must be done, or else such a Globe would be destroyed;) I say these Things are soe improbable, nay impossible, that they ought not any Longer to be mentioned or believed by us. Wherefore, for a Conclusion, I shall only acquaint, (as I am fully perswaded,) That the Sense of the *reall* Influence of the Sun; has, for want of the aforesaid, or a [p. 44] Like Examination, induced soe many, Wise and Learned, to believe; that there was in Like manner, a respective Influence peculiar *to*, and communicable *from*, the rest of the Planetts; and thereupon, the Moon (contrary to Nature and Reason) has, from time to time, been allowed, to be the Governess of the Waters of our Earth; and, (with the Rest of the Seaven,) foolishly and unreasonably, deemed the Authors, of the good or ill fate of its Inhabitants. But should we conclude, because this vast Globe of Fire (Viz.t) the *Sun*, (the proportion of which to the rest of the Planets is, in some sort, in the said Diagram exhibited) dos, at a great Distance, by its heat and Sublimations, Sensibly warme and Influence us; That therefore others, that are not only very Small, as by the said Scheme appeareth, but also Earthly and Cold, have Influence likewise; & do thereby affect us, at a like, or far *greater* distance; I say thus to conclude, would be, as if a Man should pretend to warm his hands at an Iron Bullet that is frigidly Cold, because he may do it, at one that is red hot; Which to Suppose, or effect, would be contrary to *Nature*, *sense*, *Reason*, and *Experience*; and therefore I hope this may Sufficiently evidence, that the waters in the Ocean cannot be Influenced, or agitated, by the Moon or any other, cold, earthly Plannett, or by any naturall cause or causes whatsoever; other then what is before and will be herein after Assigned.

And now I come to what was before proposed, (Namely) to Shew, That there is noe such *Correspondence*, between the Consequentiall *Tides*, and the Motion of the

Moon, as is Vulgarly pretended. And in the first place, tis to be noted, That because the dayly difference of the Moon's comeing to the South, (as by the common Tide Tables appeareth)[72] is 48. Minutes. Therefore they pretend, That the Dayly Difference from one Tide to another, is the same likewise—And *Secondly*, That on what point of the Compass soever the Moon is; when it is full Sea in any port; on her comeing to the *Same* point, and its *opposite*; She causeth full Sea [p. 45] again, in the same place, at all Times Both which I have found by observations to be utterly false; For on my apprehending the Moon, to have no power over the Waters, I kept the following Regester[73], of the Ebbing and Flowing of the Tides, (with some intermissions,) for two or three years Successively; By which I found the true *difference*, and Nature of their Flowing: Observing, with all, the point of the compas, or Azimuth, on which the Moon was at such times.

And, as to the *first* I allwayes found, That about Three Dayes after the new and Full moon, (it being then the greatest Spring=Tides; and not at the New and full, as is commonly reckoned) the dayly difference in their comeing (unless the Windes interupted them,) to be about 28. or 30. *Minutes* And, about three Days after the first and Last Quarter, (it being then the greatest Neap Tides) I allwayes found the dayly difference to be 90. 95. or 100. *Minutes* Whereas, the said difference in the Motion of the Moon, is 48. *Minutes* only: By which 'tis Manifest, That there is noe such Correspondence in their Motion, as is Vulgarly reputed: Neither can, as I humbly conceive, the said unequall differences, in the comeing or flowing of the Tides, be any ways accounted for, in the Motion of the moon: The inequality of which will more plainly appear by the following Register and by the Tide=Table from thence Æquated, and deduced.

A REGISTER

(48)

A Register showing the Time of Highwater at Weymouth Bridge; kept by this Author, Anno. 1700. &c. To which is added, The time of the New, and Full, and other States of the Moon: to be compared therewith. Being the same, from whence, the following Table was Æquated, and deduced.

1700.		Moons State	Full Sea		morning Afternoon	1700		Moons State	Full Sea		Morn. Afterne.
			Ho.	Min.					Ho.	Min.	
Aprill	15		10	30	mo.	May	18		. .	. .	
	15	D.8. mor.	. . .				19		. .		
	16		1	13	af.		20		6	45	af.
	17		3	00	af.		21	☉.5. af.	7	30	af.
	18		4	25	af.		22		. .		. .
	19		5	33	af		23		8	25	mo.
	20		6	28	af		24		9	00	mo
	21		7	15	af		25		9	19	mo
	22	☉.7. m.	7	37	af		26		. .		. .
	23		7	52	mo.		27		10	07	mo
	24		8	37	mo.		28		. .		. .
	25		9	57	mo.		29	C.8. af.	. .		. .
	26		9	55	mo.		30		12	08	af.
	27		10	11	mo.		31		1	36	af.
	28		. . .		. .	June	1		3	04	af.
	29		11	03	mo.		2		. .		. .
	30	C.2. m.	11	53	mo.		3		3	00	af
May	1		1	07	af.		4		5	55	af
	2		2	53	af.		5		6	42	af
	3		. .		. .		6	●.2. mo	7	30	af
	4		5	0	af.		7		8	00	mo.
	5		5	58	af.		8		9	00	mo
	6		6	30	af		9		. .		. .
	7	●.6. af	7	12	af		10		11	00	mo
	8		7	52	af		11				
	9		. .		. .		12	D.5. m.			
	10		8	53	mo		13		12	37	af
	11		9	15	mo		14		2	00	af
	12		. .		. .		15				
	13		11	00	mo.		16				
	14	D.2. af.	11	55	mo.		17				
	15		1	00	af		18		5	54	af
	16		2	26	af		19		6	45	af
	17		3	48	af		20	☉.6. m.	7	30	af

A Register of the Tides.

1700	Moons State	Full Sea Ho	min	Morning Afternoon	1700	Moons State	Full Sea Ho	mi	Mor: Afte noon
June 20		7	50	af	Aug: 5		8	16	mo
21		7	50	mo	6		9	15	me
22		8	11	me	7		9	45	me
22		8	30	af	8		...	...	..
23		...	...	..	9				
24		9	15	mo	10	D.11.m	11	28	me
25		...	...	..	11				
26		10	22	mo	12				
27		11	00	me	13				
28	☾.10.m	...	...	..	14		4	6	af
29		12	08	af	15		5	00	af
30		1	15	af	16		5	45	af
July 1		...	...	..	17		6	50	af
2		4	25	af	18	☉.1.af			
3		...	...	..	19				
4		6	08	af	20		8	15	af
5	●.10.m	6	48	mo	21		8	30	me
6		7	55	me	22		9	00	me
6		8	00	af	22		9	28	af
7		9	00	me	23		9	37	me
7		8	42	af	24		9	45	me
8		9	27	mo	25				
9		10	08	mo	26	☾ 7.m			
10		10	45	mo	27		12	00	noon
11		11	58	mo	28		1	45	af
12	D.1.m	..		..	29		3	15	af
13		...	...	..	30				
14		...	...	..	31		3	50	af
15		2	08	af	Sept. 1		...	...	..
16		...	...	..	2	●.1.m	7	28	mo
17		5	25	af	3		8	19	me
18		...	...	..	4		9	00	me
19	☉.9.n	6	55	af	5		9	33	me
20		7	15	mo	6		10	00	me
Omitted till August.					7		10	33	me
Aug: 1		4	51	af	8		11	15	me
2		6	00	me	9	D.1.m	12	00	noon
2		6	22	af	10		12	13	af
3		6	51	mo	11				
3	●.3.af	7	23	af	12		3	30	af
4		7	37	af	13		4	52	af

A Register of the Tides

1700		Moons State	Full Sea Ho	Full Sea mi:	Morning After:
Sept:	14		5	45	af
	15		..	..	..
	16		..	..	..
	17	⊙ 5. mo:	..	..	..
	18		8	00	af.
	½. 19		8	23	mo:
	20		9	00	mo
	21		9	30	mo
	22		..	..	..
	23		..	..	..
	24	☾.6. nig:	..	..	..
	25		11	53	mo
	26		1	37	af
	27		3	19	af
	28		..	..	..
	29		..	..	..
	30		..	..	..
Octob:	½. 1	●. 10. mo.	7	13	mo
	2		8	07	mo
	3		8	45	mo
Omitted untill July 1701.					
1701.					
July	3		2	00	af.
	4		2	49	af
	5		4	10	af
	6		5	20	af
	7		6	15	af.
	8	⊙.10. aft:	7	00	af
	9		..	..	..
	10		..	..	..
	½. 11		8	47	Mo
	12		9	12	mo
	13		..	..	..
	14		10		mo
	15		10	30	mo
	16		11	05	mo
	17	☾.2. mor:	11	45	mo
	18		12	15	af
	19		1	45	af.
	20		..	..	..
	21		4	30	af
	22		5	45	af
	23		6	30	af
	24	●.9. mor	7	25	af
	½ 25		7	45	mo
	{ 26		9	00	mo.
	{ 26		9	00	Af
	27		..	..	..
	28		10	07	mo
	29		10	53	mo
	30		12	00	noc.
	31	☽.1. mor.	12	30	Af

1700		Moons State	Full Sea Ho.	Full Sea mi:	Morn: After:
Augu:	1		1	17	af:
	2		2	15	af
	3		..	..	..
	4		5	17	af
	6		6	45	af.
	7	⊙ 5. aft:	..	..	..
	½ 9		8	23	mo
	10		..	..	..
	11		9	22	mo
	12		10	00	mo
	13		10	30	mo
	14		10	50	mo
	15	☾.6. af	11	30	mo
	16		12	30	af
	17		..	..	..
	18		3	00	af
	19		..	..	..
	20		5	30	af
	21		6	21	af
	22	●. 5. aft:	7	05	af
	½ 23		7	45	mo
	24		..	..	..
	25		9	24	mo
	26		..	..	..
	27		10	32	mo
	28		11	10	mo
	29	☽.10 mor.	11	42	mo
	30		12	36	af
	31		..	..	..
Sept.	1		3	25	af
	2		..	..	..
	3		..	..	..
	4		..	..	..
	½ 5		7	00	mo
	6	⊙.5.m.	..	..	..
	7		..	..	..
	8		8	30	mo
	9		8	53	mo
	10		..	..	..
	11		9	41	mo
	12		10	12	mo
	13		10	30	mo
	14	☾.7.m.	..	..	..
	15		12	13	af
	16		1	53	af
	17		..	..	..
	18		5	37	af
	19		6	15	af
	20		..	..	..
	21	●.2 m.	..	..	..
	½ 22		8	15	mo

A Register of the Tides

1701	Moons State	Full Sea Ho	Full Sea mi	Morning Afternoon	1702	Moons State	Full Sea Ho	Full Sea mi	Morning Afternoon
Sept. 23		9	06	mo	July 30				
24		9	30	mo	31		9	15	mo
25		10	00	mo	Augu. 1		10	00	m
26		11	04	mo	2				
27	D.8.n	11	30	mo	3		10	40	mo
28		...	...	...	4	☾.5 af	11	15	mo
29		1	15	af	5				
Omitted till June 1702					6		12	55	af
					7				
June 25		4	37	af	8		5	50	af
26		5	32	af	9				
27		5	57	af	10		6	05	af
28	⊙10.m	...	...		11				
29		8	00	af	12	●.9.m	7	12	af
♄ 30		8	03	mo	13				
July 1		9	17	mo	♄ 14		8	52	mo
2					15		9	30	mo
3		...			16				
4		11	00	mo	17		10	35	m
5	☾ 11 nig	...	...		18				
6		11	51	mo	19	D.5.m	12	05	af
7		12	45	af	20				
8		2	00	af	21		2	50	af
9		5	30	af	22		5	57	af
10		4	50	af	23				
12					24		6	55	af
13	●.10. nig	7	00	af	25		6	49	af
♄ 14		...			♄ 26	O.8.m	7	15	mo
15		8	10	mo	27		...	...	
16					28		8	25	mo
17		9	40	mo	29		...	...	...
18		10	15	mo	30		...	...	
19		...			31		9	40	mo
20	D.10 nig	11	50	mo	Sept. 1				
21		...			2		10	45	mo
22					3	☾.11 mo	...		
23		2	40	af	4		11	45	m
24		3	21	af	5		12	30	af
26		...			6				
27	⊙.8 nig	7	20	af	7		4	30	af
28		...	...	...	8		5	55	af
♄ 29		3	15	mo	9		6	22	af

A Register of the Tides.

1702	Moons state	Full Sea Ho	Full Sea mi	Morn Aftern
Sept: 10	●.6.n.	7	00	af
½ 11		7	55	mo
12		8	21	mo
13		9	08	mo
14		9	45	mo
15		10	30	mo
16		11	10	mo
17	☽.10.m.	..	..	
18		2	30	Af
19		..	..	
20		..	..	
21		..	..	
22		5	45	af
23		..	..	
24	○.11.ni	6	55	af
½ 25		7	45	mo
26		8	15	mo
27		..	..	
28		..	..	
29		..	..	
30		10	00	mo
Octo: 1		..	..	
2	☾.4.m	10	30	mo
3		11	10	mo
4		11	00	mo
It flowd twice in 4 Hours 5		2	45	Aft
6		4	25	af
7		5	06	af
8		6	00	af
½ 9		6	30	mo
10	●.3.m.	7	30	mo
11		..	..	
12		9	00	mo
13		9	30	mo
14		10	22	mo
15		10	30	mo
16	☽.8.ni	11	15	mo
17		12	15	af
18		..	..	
19		3	15	af
20		3	00	af
21		5	30	af
½ 22		6	00	mo

1702	Moons State	Full Sea Ho	Full Sea mi	Morn Aftern
Octo 22 Full Sea at Abbotsbury at		10	45	mo
23		..	..	
24	○.5.ni	7	30	mo
25		..	..	
26		8	08	mo
27		..	..	
28		9	04	mo
29		..	..	
30		9	42	mo
In the West Bay near the Isle of Portland at				
30		10	00	mo
31		10	48	mo
Novem: 2	☾.7.ni	12	37	af
3		1	30	af
4		3	30	af
5		4	07	af
6		5	30	af
7		..	..	
8	●.1.af	..	..	
½ 9		8	15	mo
10		..	..	
11		9	15	mo
12		..	..	
13		10	25	mo
14		11	08	mo
15	☽.9.m	..	..	
16		12	15	af
17		2	05	af
18		3	35	af
19		4	08	af
20		5	15	af
21		..	..	
22		..	..	
½ 23	○.1.af	7	04	mo
Abbotsbury 23		10	25	mo
24		11	35	mo
24		..	..	
25		8	15	mo
26		..	..	
27		9	08	mo
28		10	16	mo

A Register of the Tides.

1702	Moons State	Full Sea Ho	mi	Morn/Aftern
Novem: 29		..	..	..
30		10	53	me
Decem 1	C.8.m	12	00	noo
2		12	45	af
3		..	..	
4		..	..	
5		4	30	af
6		..	..	
7	●.11.n	7	30	mo
8		8	00	mo
9		..	..	
10		9	30	mo
11		10	00	mo
12		10	30	mo
13				
14		11	15	mo
15	D.2.m	12	15	af
16		1	15	af
17				
18		3	30	af
19		4	10	af
23	O.7.m	..	..	
24		8	00	mo
25				
26		9	8	mo
27				
28		9	52	mo
29		10	45	mo
30	C.7.ni	..	..	
31		12	15	af
1703/3 Janu 1		1	15	af
2		2	30	af
3		..	..	
4		4	30	af
5		..	..	
6	●.9.m	5	23	mo
7				
8		8	00	mo
9		9	20	mo
10				
11		10	16	mo
12				
13	D.9.n	11	00	mo
14				
15		12	02	af
16		1	10	af
17		..	..	
18		4	10	af

1702/3	Moons State	Full Sea Ho	mi	Morn/Aftern
Janua 19		..	..	..
20		5	15	af
21	O.11.n	..	..	
22		7	17	me
23		7	45	me
24		..	..	
25		9	04	me
26		10	00	mo
27		11	34	me
28				
29	C.5.m	12	15	af
30		..	..	
31		..	..	
Feb 1		3	52	af

Omitted till March following

March 24		8	52	mo
1703 25		9	15	mo
26		10	17	mo
27		10	47	mo
28	C.5.n	..	..	
29		12	00	me
30		1	30	af
31		3	15	af
Apri 1		..	..	
2		5	40	af
3		6	30	af
4				
5	●.4.m	..	..	
6		7	48	me
7		8	30	mo
8		..	..	
9		9	12	me
10		9	30	mo
11		..	..	
12		10	45	mo
13	D.8.m	11	45	mo
14		12	30	af
15		2	50	af
16		4	04	af
17		5	15	af
18				
19		7	00	af
20	O.10	..	..	
21	mo	8	00	mo
22		8	30	mo
23		9	52	mo

A Register of the Tides.

1705		Moons state	Full Sea Ho	mi	Morn After			Moons State	Full Sea Ho	mi	Morn After
Aprill	24		10	17	mo	June	7		9	15	me
	25		--	--	--		8		9	53	mo
	26		11	54	mo		9				
	27	C.1.m	12	50	af		10		11	15	mo
	28		1	54	af		11	D.10.m	12	19	af
	29		3	50	af		12		1	30	af
	30		4	25	af		13		--	--	--
May	1		5	50	af		14		3	45	af
	2		--	--	--		15		4	57	af
	3		--	--	--		16		6	00	af
	4	●.9.n	7	29	af		17		7	04	af
	5		8	15	af	½	18	○.1.m	7	45	mo
½	6		8	15	mo		19		--	--	--
	7		--	--	--		20		--	--	--
	8		9	02	mo		21		9	51	mo
	9		--	--	--		22		10	30	mo
	10		10	10	mo		23		11	15	mo
	11		--	--	--		24	C.10. nig	12	00	me
	12	D.11.n	11	15	mo		25		12	45	neo
	13		--	--	--		26		1	15	af
	14		2	06	af		27		--	--	--
	15		3	27	af		28		3	31	af
	16		--	--	--		29		4	30	af
	17		5	30	af		30		5	35	af
	18		--	--	--	July	1				
	19	○.7.n	7	18	af		2		6	45	af
	20		7	43	af		3	●.2.m	7	50	af
	21		--	--	--		4		--	--	--
½	22		9	00	mo	½	5		8	45	mo
	23		--	--	--		6		9	19	mo
	24		10	51	mo		7		9	45	mo
	25		--	--	--		8		10	30	mo
	26	C.11.m	--	--	--		9		11	54	mo
	27		11	48	mo		10	D.5.m	--	--	--
	28		2	15	af		11		--	--	--
	29		3	30	af		12		1	45	af
	30		--	--	--		13		5	08	af
	31		5	36	af		14		4	50	af
June	1		6	08	af		15		--	--	--
	2		6	45	af		16		6	54	af
	3	●.11.m	7	15	af		17	○.5.m	7	45	af
	4		7	45	af		18		--	--	--
½	5		8	12	mo		19		8	58	af
	6										

A Register of the Tides.

1705		Moons State	Full Sea		Morn Aft
			Ho	mi	
July	20		9	30	mo
	21		10	04	mo
	22		10	45	mo
	23		11	30	mo
	24	5 1 m	12	02	noo
	25				
	26		1	12	af
	27		2	15	af
	28		3	30	af

Note, The ½. shews that the Difference must be recconed but for halfe a Day.

Note, Also, That this great difference, and somtimes the very small Difference, that is in the Flowing of the Tides, is accidentally occasioned, by the Suddain Shifting of the Winds; By means of which the Tides aither haltened, or retarded, in their Motion.

Moon's Age	Full Sea Ho.	Full Sea Min.	Dayly Differ. minu	The Difference in the two Tables; Before and After each other — Minuts	Moons Age	Full Sea Ho.	Full Sea mi	Dayly Differ. minu
00	15	12	33	00	0	15	12	
1	16	12 32	32	16.b	1	16	12 48	48
2	17	1 01	30	34.b	2	17	1 36	48
3	18	1 30	29	54.b	3	18	2 24	48
4	19	2 01	31	71.b	4	19	3 12	48
5	20	2 35	34	85.b	5	20	4 00	48
6	21	3 13	38	95.b	6	21	4 48	48
7	22	3 57	44	99.b	7	22	5 36	48
8	23	4 52	55	92.b	8	23	6 24	48
9	24	6 02	70	70.b	9	24	7 12	48
10	25	7 37	95	23.b	10	25	8 00	48
11	26	8 52	75	04.a	11	26	8 48	48
12	27	9 54	62	18.a	12	27	9 36	48
13	28	10 45	51	21.a	13	28	10 24	48
14	29	11 27	42	15.a	14	29	11 12	48
15	30	12	33	00	15	30	12	48

A Table of the Flowing of the Tides as it was found by observations made Anᵒ 1700. 1701 &c. in Weymouth Harbour by W. H.

A Table of the Flowing of the Tides according to the Vulgar account; or as if they were Governed by the moon.

The difference or maner of their Flowing — — described by Lines from Spring, to Neap Tides.

Days of the Moon's Age.

D 2½ 3 4 5 6 7 8 9 10 E
At Springs Tides
At Neap Tides
F 2½ 3 4 5 6 7 8 9 10 G

Days of the moons Age.

How the?

By the Line FG. is shewn, (to wit) Tides would flow, equally) if they were Governed by the moon.

By the Line DE. is shewn, how they do flow; as tis found by Observation, and experience: By which it appears, That the Difference in their Flowing; is in an Unequall, Logarithmeticall proportion; and not in an Equall progressive difference; as has been Vulgarly accounted. Which the dayly difference in the two last Tables, ——

from whence

By the Line F.G. is shewn, How the *Tides* would flow (to wit) equally) if they were Governed by the *Moon*

By the Line DE. is shewn, how they do flow; as 'tis found by Observation, and experience: By which it appears, That the *Difference* in their Flowing; is in an Unequall, *Logarithmeticall* proportion; and not in an *Equall progressive* difference; as has been Vulgarly accounted. Which the dayly difference in the two last Tables, [p. 55] from whence these Lines were Extracted, will also evidence; And will more fully appear by the following Explanation.

By these Tables, it appeareth, That when the *Moon* is 4. or 19. Dayes old, 'tis Full=Sea, according to truth or Observation, at 2. of the Clock and 01. *Minute*. But the Same Dayes, by the Vulgar account, at 3. of the Clock and 12. *Min*: *which is 75. Min*: *or 1. hour and 15. minuites Difference; as by* the Collumn between the two Tables appeareth; and so of the rest. And on the other hand, it is also to be Noted; That 3. or 4. Dayes *after* the New, and Full Moon; the Dayly difference in the comeing of the Tides, is about 28. or 30 *Minuites*, But about three Dayes *after* the first and last Quarter, 'tis found to be 90. 94. or 100. Minuites; as is before observed. To which the common Tide= Tables, ought to be conformed; That the disappointments, that may happen, for want of knowing the true State of the Tides, may be prevented. And that this is the *naturall* Motion of the waters, is the more certain; because a great part, of these Observations, were made in the year 1700. when by reason of the Dryness of that Sumer, the naturall Motion of the Sea, was not interupted by Windes, as usuall. Nor is it here, at any time, by any River: For this is very small, and riseth, but about two miles distant, from the sea= Coast.[74]

And as to the *Second* part of the Assertion, wherein 'tis Vulgarly Said, That on what point of the Compass Soever, the moon dos make ffull Sea, in any Harbour; on her coming to the *Same* point, and its *Oposite*, she there maketh full Sea again; This (as is said,) [p. 56] I have frequently found to be false, by two or three points at a time. But that which dos fully confute the said Error; is, That whensoever the *Moon* is said to be in the Latter Degrees of *Gemini*, and in a great part of the Signe *Cancer*, she dos then pass, from the *East*, to the *West* point of the horizon; in about 8. or 9. Houres; and from *West* to *East* again; in about 15 or 16 Hours; in our Lattitude; and farther South, she dos the like in 4. or 5; and in 19. or 20. hours: Now according to this Account, the coming of the Tides at such times, would be but 4. or 5. houres distance **in** one Tide; and 8. or 9. Ho: distance in another; which dayly Experience dos fully confute. The *comon* distance in the coming of the Tides, being allwayes between 12: and 13. Hours.

But tho' I observed this difference, in their *inter=meditate* Motions: Yet I found no such, in their Monthly *Revolutions*; For I alwayes Observed, That whensoever the Moon was about 3. Dayes past the full, and as many past the Change, (falsely so-called for she's at the Full, to a trifle, equally enlightned, (unless when Eclipsed,) at all times,) it was constantly the *greatest* Spring=tides; unless violent Windes interrupt=rupted the

Same; Now this Discord in one Part of their Motions, and conformity in the other, I confess, was for some time, the Subject of my thoughts; Till at last I considered, That the Waters in the Ocean; were, by the pulsing of the Earth, thrust farther off, from the Earth's Center; than otherwise it would be; From whence I Conceived That Such an Elation might naturally bring the Earth out of its *Equilibrity*; (tho I *then*, nor 'till Some years after, knew or conceived the *place* or places where [p. 57] they were soe moved) and thereby cause its Diurnall Rotation; as shall herein after be further evidenced. Which haveing I say, thus conceived I farther traced its Effects; And well knowing that the Earth, and the Moon; are both Suspended in the open *Æther*, or Sublimations of the Sun; and not fixed in any Grosser matter; I thence collected, First, That the Earth's Rotation, dos, (as experience confirmeth.) Give motion to its own *Atmosphere*— Secondly, That this Atmosphere, being (as is said) Suspended in, and enclosed by the Said *Æther*, must unavoidably give motion thereunto: And lastly, the said *Æther* being moved as aforesaid, and also touching the Body of the Moon, and she haveing no Rotation of her own (the Earth being the Center of her Orbit;) I say this Æther might, by the contiguty of its Matter, move, or draw round, the Body of the Moon, after the Earth; and thereby cause her, to make one Revolution, in the time that the Earth maketh 27 & $\frac{7}{24}$.[75]

Now seeing what is herein advanced, is, (I presume,) in some respects wholy new or at least, but rarely treated of; It cannot therefore be expected, I should confirm it by the Testomony of former Writers; Yet that I may not be found arbitrary in my Assertions; Will therefore Indeavour to prove them, by an easie and familliar Experiment: which in deed is what has too frequently been wanting in those who have Written on things of this Nature. As for Instance, one Author affirms,[76] *That the Earth at the* Deluge, *was broken into pieces, and the four Quarteres thereof,* raised *by the Tilting up of some of those peices.* Another Supposeth and Asserts, other things; [p. 58] But neither of them gives us any Experiment, to prove their Propositions; and therefore they may be, all of them, as well false, as true; For Speculative Notions, without practicall Experiments, or Visible Phænomena's, are not Sufficient foundations, for an *Hypothesis* of this kinde. I shall therefore by the following Experiment prove or illustrate the particulars before mentioned. And in order thereunto, it is to be Observed,

> *That if any* Gross Body, *be moved round, in any* Fluid Matter; *it will not only, give a like Motion, to the Said matter; But also move,* any other *Gross Body, swimming on or Suspended in the Same fluid; tho' it be posited at a Vast* distance *from the Body* first *moved.*

Now that this is true, will visibly appear, by puting a Round Tubb, or other like Vessell, into a Pond or Cistern of Water; Wherein if you move it circularly; or turn it but a Little time on its Center, as a Mill=Stone is (to wit) parellel to the Horizon; You will See the *whole* quantity of Water; as also my thing floating tho' at a distance in it, gradually, & naturally, to move Round therewith—In which, the Revolution of the Body so drawn round; will be in a just proportion to its distance from the Body first moved.[77] Which so exactly agrees, not only with the proposition above mentioned, But also with the worthy *Keplers* late Discovery (Viz.[t]) "*That the time Spent in The* [*p. 59*] *Revolution of the Planetts, are in a Fixt, certain proportion, to their Distance from the Center of their Orbit;* as that it even Demonstrates the same.[78] Now that the Earth has a Diurnall Rotation, is (especially of late years,) So generally granted; that I shall only refer you to the Post=Script, hereunto annexed, for its Confirmation. and

Secondly, That the Diurnall Rotation of the Earth, dos Give Motion to its *Atmosphere*, is evident; in that its Atmosphere, dos always turn round, and accompany the Earth, in its Rotation; and that even in all parts of the world; unless about 30. degr:[s] on each side the *Equinoction*: and in that part too, in the Continents; where the Hills and Mountains Stand up in the *Æther*, to further the said Motion. So that the *Trade=Winds*, which are those that do not quite keep an even pace with the Earth, are only found on the Smooth Ocean, between *Africa*, and *America*, and thence to *Asia*; where, we know, there can be no Mountains Standing up, to interupt the Said Windes: or to Cause the *Atmosphere*, to turn with; and accompanny the Earth, as it dos at Land: And if you would have this turning of the Earth, in its Atmosphere; (whence the said windes are produced,) plainly represented, and illustrated by an Experiment; It is but to draw a peice of Board, or the Like, on the Surface of a Cistern of Barly when the water is first fill'd up, about 2. or 3. Inches above the Corne; or on any Shallow water, haveing Dust or Chaft Strawed thereon; And you may see the Floating Dust, turn in, on each side, as you draw it along; just as the Said windes do, on each Side the Æquinoctiall Line: [*p. 60*] And (if drawn from *West*, to *East*,) the waters will turn in on the Same points likewise; And by the uncertain or irregular rufflings, and Calmness, of the Wateres; at and near the End of the said Board; will in like manner be plainly Shewn; the Calms, and intermitting Concussions, of the windes; on the West Side of *Africa*, and else where; called the *Tornedoes*. And what is more, if Two Boards, be cut in the Forme of the two Mapps, (or rather, only of the Forme and breadth, of that part of them which is between the two *Tropicks*) and fastned, or joyned by two or 3 Wires, standing up, arching over the waters; I say you will, in drawing them as aforesaid, evedently see the Cause, Why the said Windes turns in, on the South side of the said Line, on the Coast of *India*; and not on the North: As also the constant Calms, that are found in those and other parts of the World. All which I have with pleasure beheld by the Said Experiment. And as the turning of the Earth, is the Cause of the said windes; so the very Being of these Winds, dos plainly Demonstrate the said *Rotation* of the Earth; For if the Earth did not so turn; it would be naturally impossible, that those Windes should be Produced.[79]

Now seeing it is evident that the Rotation of the Earth, dos so move its *Atmosphere*, (or at least that part which toucheth, or is nearest to the Earth,) as to cause it almost in all places, to keep an even pace with its own Body: And seeing it cannot be denied, but that the fluid *Æther* or Sublimations of the Sun, dos touch and incompas the Earth, and its Atmosphere; and also the Moon & Her's, as by Fig: 10th appeareth: And seeing also, in the said Experiment, [p. 61] the fluid matter that toucheth, or is nearest to the Body *first* moved; dos almost accompany ye Same; Whilst that at a greater distance, will in proportion, move more Slowly: **How** can it be denied from the conformity of those Motions, and the contiguity of the Said Ætherall Matter; But that the Earth, by its Diurnall Rotation, dos move its *Atmosphere*; and that this Atmosphere, dos move the *Æther*; and consequently That the said *Æther*, dos give motion to the Body of the *Moon*; altho', by reason of her Distance, her Motion is about 27. times Slower than that of the Earth; as is before observed—So that by what has been Said, I hope it is evident; That the Moon is so farr, from being the *Cause* of the Flowing of the Tides; as that instead thereof, the lifting up of the Ocean, (whence the Tides are produced, as will herein after be shewn,) is the originall Cause of the Motion of the Earth; And consequently, that the Rotation of the Earth, is the very *Efficient Cause*, of the Motion of the Moon: Which I presume may Suffice to Shew; why the *Spring*=Tides do agree with the full and Change thereof; as was before observed. I Shall now proceed to Some other *Pænomena's* that are Vulgarly said to be caused by the Moons pretended *Influence*: And herein shall instance only in three Sorts, because I conceive, that most, if not all the rest, may be thereunto referred, and the *First* shall be of *Lunacy* which by reason of its revolving with the Motion of the moon, as is pretended, may seem to be governed by her, and yet notwithstanding is not. The *Second* sort Shall be of Such as do really depend upon the Moon for its Cause, in respects of its *Light*, but not otherwise. And my *third* and last, shall be of Such as are not reall, but only imaginary or at best [p. 62] but uncertain Phenomena's.

    And First, as to that humane Infirmity called *Lunacy*[80] we all know the Earth is the Mother or Originall of all living Creatures, man himself not excepted. And inasmuch as all Creatures, do generally resemble or partake of the nature of the Dam or Root, from whence they had their Being, whether they be produced of the Same or of another *Species* (as is manifest in the Smaller Animalls, bred by the Heat, and sweat, of Doggs and Sheep &c. The pulse or circulation of the Bloud, being the very same, proportionably, in the *lesser*, as in the *greater* that produceth them.) And Seeing there is in such persons at such times, as they are so affected; a greater flowing, or a different Motion, or Circulation of the Blood, than at others Seasons, or States of the Tides; it is but naturall and reasonable to conclude, That the cause of this Infirmity dos proceed from the said superabounding of the Blood, or of its motion, in the *Lesser* Animalls; at the same time, and in conformity to, the Like grand or Monthly fluxing of the Waters, in the *greater* Animall; I mean to the pulsing or flowing of the Ocean, at or near, the new and full Moon; upon which the coming of this Distemper, is said to depend: And, if soe, whether this inference; be not more reasonable & Naturall; than to allow the

Moon; at so many thousand Miles distance; to be the Cause of such a *Phenomena* in humane Nature; I leave the Reader to Dertermine; as also of all others, that may be thereunto referred.

The *Second*, is the Monthly increasing & decreasing the Flesh of Some Fowles; (and perhaps other Animalls;) which indeed at First, would seem an irresistable Argument; But [p. 63] when we consider That the *Heron*, the *Owle*, &c. (being those in which the *Phænomena* is found) are such Creatures as doe cheifly seek their prey by night; Tis no Mistery to conceive, that the *Light* of the Moon, when at and near, the full; should contribute to their more plentifull Feeding; as also That the *darkness* of the Nights, at the Change, should prevent them therein: From which vicissitudes, of Light and Darkness, their Bodyes must gradually, be increased or diminished accordingly: But nevertheless, the moon cannot be the Cause of this Phænomena; (other than by its reflecting the Light of the Sun upon the Earth;) for if soe, it may be as well said, that the Candle threads the Needle, and not the Man; or the Sun dos Catch the *Hare*, and not the *Dogg*; because they doe, in like manner; comunicate their Lights to the performance of these respective Actions.

The *third* and last Instance is concerning the Times and Seasons, that are esteemed propitious for letting Blood, cutting haire, killing Swine, and the like; to which, my Answear in generall is, That I presume they are only traditionall and groundless Imaginations; derived to us from the Antients, and not reall Phænomena's in Nature: And that the first is such, is evident from experience; For 'tis well known, that 'tis as Safe Letting Blood, when the moon is in the Sign *Gemini*, or in any other such fabulous* position wherein you propose to Bleed; as if she was opposite thereunto; or at any other distance whatsoever. But however for people to observe the full, and Change, in cutting of Hair, and the Like. I confess; may be very usefull; (tho' not in respect of the Moons Influence;) Because 'tis evident [p. 64] that the Constant, and regular cutting, and pruning of Trees, Herbs or other Vegetables, (which is the same with the Hair in an Animall) will Cause them, to take Root and flourish the Better: And in this Respects, a like *regular*, or constant observing of that, or any other Season, for cutting hair; may further its Growth and Increase; whereas by a careless Neglect; the Effect may be otherwise. But tho' this be So Yet inasmuch as we all know, that those Vegetables are produced, and nourished; only by the heat, and humidity of the Earth, (intermingled with the accidentall Sublimations of the Sun:) and knowing tis the Same with the Hair, in respect of the humidity and heat of the Animalls that do produce it; There can therefore be no Room for the *Moon's* pretended Influence: And as

---

* *I call them* Fabulous *because their very being, as well as their pretended Natures, are only feigned, and imaginary Institutions, or Astrologicall Impositions, and groundless Machinations: And were it not soe yet according to the Copernican System, now approved; it may as well be said, that the Clouds, when moved in the Air, doe pass from one sign to another, as it may be the Moon, or any the rest of the Planetts (as could easily be demonstrated.)*

Little Occasion, for observing her 30. Days Revolution: For the event will be the *Same*, if you Observe, but 5. or 6. and twenty; or 7. or 8. and forty, dayes Distance, in Such Cutting or pruning: as if you ahould keep exactly to the Revolution aforesaid. And as for the Seasons vulgarly appropriated for killing of *Swine*; I must acknowledge, I have soe little Esteem for *Lunaism*, that I leave it to Such as are Judicious therein, to finde out a Reason; Why one State of the Moon, should be more propitious, than another, for killing *that* Creature; and not for Cows, Oxen, Sheep, or other Animalls.

There is one other Argument that I would have omitted, had it not been lately urged against me; (viz.t) The Moon (says they) dos certainly govern yᵉ Tides, because it Governs yᵉ Fluxions in the Body of *Venus*, and her offspring: To which I answear, That 'tis rather an Argument to the contrary; And that, because tis Manifest, that 'tis only when Nature disposeth thereunto: For if you consult Madam—she can tell you that in 20. of them it will be found at twenty Severall Times, or States of the Moon; and not only at the new and full as is pretended: And that even [p. 65] in those whom you would deem to be one and the Same *Constellation* or *Complexion*.

Thus have I attempted to Shew the invalidity, of all the old and vulgar Arguments, now accruing to memory, which are offered, to prove, That the *Moon* has *Influence*, over the Terrestriall waters; or over the Animalls that were thence produced: but yet I finde one other of a Fresher Date, which I cannot pass; For altho' Some of these Authers seem to lay aside the old Notion of the Moon's Governing yᵉ Waters by a Magnetick Influence;[81] (which I was pleased to hear, in expectation, that between the two Stools, both Errors would have fallen to yᵉ Ground) yet instead thereof, they affirme the same motion, to be produced or effected, by the pretended *Squeesing* or *pressing* power of the Moon,[82] which they would demonstrate by the following Observation; (viz.ᵗ) "*That heavy Bodyes, incumbent on the Center of their Gravity; the* "*nearer they approach thereunto; the more they Gravitate:* From whence 'tis inferred, That "when the Moon is in her *Perigæum*, or nearest to yᵉ Earth; She dos press harder, (I "suppose on the Æther) than when in her *Apogæum*, or greatest Distance from it: And "soe consequently, according as the Degrees of Pressure is made; the *Tides* are either "Increased or diminished.

Now that the Bodyes soe approaching, may gravitate as aforesaid, when removed from their naturall position; I will not dispute; for 'tis very probable, that if one of the Rocks of the Moon at N. (see Fig: 11ᵗʰ.)[83] could be removed thence towards the Earth, it would before it came to B. have the *Same* Inclination to the Earth, as to the Moon; and soe on the Contrary; And therefore at such a Distance it would precipitate more *Slowly* than in its nearer Approaches, unless when arrived to our [p. 66] *Atmosphere* where its grossness, would doubtless somewhat retard its motion: But tho' this Observation may be True, yet it dos not follow, that the Inference thence drawn, is soe likewise, For we See in other Examples, That *Erroneous Notions* may be derived from reall Phænomenas; As for Example, in the *Torescellian* Experiment.[84] The Pressure of a

Collumn of Air, being the Assigned Cause of the Rising and Sinking of the Mercury in the Tube; some will therefore Infer, That the Motion of the *Thorax*, and consequently our Lives, are owing to the Like pressure;[85] But that this is a groundless supposition, is evident from our Liveing soe long underwater, in those Sort of diving Engines[86] Wherein the Said Collum, and its pressure, is wholy interrupted whereby, if it were So, Death would imediately ensue; But we finde by experience the Contrary.

Another like Notion is that of Monsiur[87] when he traveled with one of the *Asian* Princes; wherein he pretends, with the Said Instrument, to Measure the Height of a very high Mountaine, or rising Ground, tho' it required Severall Dayes Journey to Ascend it: In which Time, 'tis well known; that the very Change of Weather might naturally Cause a great difference, in the Hight of the Mercury; and thereby render Such a Calculation very uncertain and Erroneous—And as tis thus uncertain, soe likewise needless; For by a Chain,[88] Or the Wheell of his Coach, and a Land Quadrant, he might have [p. 67] taken the *Hypothenusa*, and the Angle at the *Base*, and thereby have obteyned his desire, to a moral perfection. From which Instances as also from others of like nature, it is evident; that Erroneous Inferences may be derived from reall *Phenomena's*, we will therefore now Enquir whether the pretended *pressure* of the moon before mentioned be not one of that number.

And in the *first* place, I humbly conceive, That such a *pressing*, or Gravitateing Power is not to be found in any of the Celestiall Orbs; Specially whilst they are naturally posited, as the moon is (tho it may be supposed of Commetts, or of Bodyes occasionally removed as aforesaid:) And my Reason is, Because if soe, whensoever either *Saturne*, *Jupiter*, or *Mars*, should approach towards, or come in opposition to the *Sun* (which is their assigned Centre of Gravity,) they would then Gravitate towards the Earth, as well as towards the *Sun* (seing the Earth is at such times always between the Sun and them; as will plainly appear by Figure the tenth; and the better if you turn up the Book, so as that *Saturn* may be at the top of the Said Figure,) Whereupon I say, if there were such a pressing Power in them, they would then cause Such great Tides, That we Should be induced to conclude; That the universall Deluge, was as soon after the Creation, as those Three Vast Bodys, (or but any two of them) came, in some proper Signe, to such an opposition of the sun; as is described in y^e Line **S.E.** in the Fig: aforesaid.[89] And if you would know what Signe that was; To be Sure the Scheme=Makers, will tell us, 'twas in Aquarius, or the water bearer; (tho' others would have it in *Gemini*), And some reconciling Naturillist, perhaps D^r: *N.*[90] will [p. 68] refer us to the *Rainbow*, for the Reason why their Gravitateing power is *now* restrained. But how improbable, and inconsistent with Experience; these conjectures are, I leave to be determined by the judicious Reader; and proceed (2dly) To the Moons causing (as is pretended) full sea, when in her *Perigæum*, or nearest distance from the Earth: And this I conceive is not in the least to be regarded neither; for if you consult but a common Almanack, you may in one years Time, find her in that position, at all States of the

Moon; as well at the *Quarters*, as at the full and Change; and consequently if *that* was the *reall* Cause; it would accordingly be spring=Tydes; some times at the Quarters of the Moon; and again, at other times, at the full and Change; which common Experience dos dayly confute. Nay what is more, if there was such a pressing power as is pretended; How the *Apogæon*, and the *Perigæon* of the moon; should both produce, one and the *Same* Effect, (as by their account it must) is a self=confuting Mistery, as great, if not greater than the former.[91]

By what has been said, it may reasonably be granted, That there is noe such Gravitateing Inclination in the moon, or any of the Planets (whilest naturally posited) as is pretended; For if twa's; what should hinder them from executing their propensity? seeing the *Commetts* can precipitate, soe near the Sun without Interuption, as to be sett on fire by it. But however, that I may not seem to begg the Thing in Question; we will Suppose there is such a Nature in them as is proposed; Yet nevertheless, I conceive it is impossible, That the Body of the Moon should soe press the *Æther*, as to cause it to produce the Ebbing and Flowing of the Ocean: And that my Reasons for Such a conception, may the better appear; I shall lay down these two Propositions or Postulata's.

"First *if two, or more Gross Bodyes (of equall Magnitude,) doe move, or precipitate, in any*
"*unconfined fluid matter;* The Quantity *of such matter, by them respectively moved, will*
"*be in a just proportion, to the* Rarity, *or* Density; *of the Fluid matter, in which they*
"*move.*

[p. 69] Therefore any such Body moving in *Æther*, or Air, will move many Thousand times Less in Quantity of those thin fluids; than if moved in Mercury or in Water.

"2.dly *If any Gross bodyes should press down, or precipitate into any fluid matter,* confined,
"or circumscribed, *(in a Tube or the Like) so close to all the sides, of what contains it,,* as
"*to permitt none to press out; such body must then gradually give motion,* to the whole
"mass *of matter therein inclosed, or Conteyned.*

Therefore (e.g.) If a Tube of Air (see Fig: 12.)[92] be 7. or 17. foot Long; and if but 5. or 6. inches be pressed down at the Top: That pressure will affect and give motion to the whole; Because the Air, or matter therein contained, will gradually seek to restore an uniformity, in its consistency.

Now to explicate these Propositions It is to be noted: *First*, That the proportion of Air to Water, is found to be, as 840. is to 1.: soe that if 840. Gall.[s] of our *Gross* Air, be condenced; twill make but one Gallon of Water; and if the Top of the *Atmosphere* (allowing it, as 'tis lately found, to be 45. Miles. high) be equated with the Bottome; 'twill require 6700. Gallons, to produce that Quantity; And as the Air is thus Thiner, and Lighter than Water, soe tis so much the more feeble and yielding. For though

Water cannot be compressed yet Air will admitt of being condenced to 60. Times Less; and also of being rarified, to 70. Times more, than its common consistency: Now seeing 'tis evident, That the Air is 6700. times thinner and Lighter than Water; we need not doubt but ther's the Same, if not a greater Disparity between the *Air* and *Æther*; and consequently the *Æther* is above thirty Six Millions of Times, Thinner, than water; and [p. 70] therefore, according to the 1ˢᵗ *Postulatum*, soe many Times the more yielding and less liable to be moved or affected, by the Motion of the Moon, or of any Gross Body pressing downe, or moveing in it.

2ˡʸ That the Bodyes of the Earth, and Moon, are both of them openly suspended in this unconfined, and as it were, infinite Space of thinn fluid, transparent *Æther*. And that it is of such a vast Extent, is evident from Fig: 10ᵗʰ before goeing (tho' we should therein suppose it to be extended, but Little farther, than from the Body of the *Sun*, to the Orbit of *saturn*, only.)

Thirdly That there is also a vast Disparity between the Body of the moon and her Orbit, and also between her Orbit and yᵉ Body of the Earth; as appears by Fig: 11ᵗʰ:[93] (Wherein I have laid them down in as just proportion as the Circumstances would admitt) From whence we may also conceive, That the Bodyes of the Moon, and the Earth, are but as it were soe many *Attomes*, if compared with the Vastness of their Distance as aforesaid. And 4ˡʸ (as is also evident by the said Diagram). That the moons *Perigeon*, or approaching towards the Earth, (to which there is of late,[94] as is said, soe much weight or Virtue ascribed;) is altogether as inconsiderable, as their Magnitudes, For, if from the 114000. Miles of their mean distance, be deducted the 5000. Miles that's allowed for the Equant, or *Eccentricity* of the Moons Orbit; there will Still remaine above 100000. Miles for their nearest distance. From whence we may justly conclude, That if all these Things be duely and impartially compared, and considered; it will appear to be as impossible, that such an *Attom* as the Moon, at soe vast a Distance, Should give Motion to such a thin yielding Medium as the ther is; and thereby, cause it to press upon another like Attom, as the Earth is, so as to give Motion to the Ocean, or its fluid part, and not thrust the whole Globe out of its Orbit; as that the Man in the Moon should hear, what I am now Saying of his antient Habitation: Especially, seeing we finde, That a Cannon Bullet in [p. 71] its Swiftest Motion, in a many thousand Times Grosser Matter, will not move a *Feather* (excluding the Motion of the common Air, made by the Explosion of the powder at the Mouth of the Cannon) perhaps half a Dozen Yards before it comes to it, much less at its first approaching (to wit) at 2. or 3. thousand yards distance.

But, if what has been said, be not yet thought Sufficient; we will Illustrate the Matter by another Example. And herein it is to be noted, as has been intimated, That the Magnitude of the Moon, is to her Orbit near as 2. is to 100. and her *Perigæon* as 5. is to 100: Therefore, according to such proportion, Lett us suppose, (as in Fig: the 9ᵗʰ at A.)[95] That a Sphere of 2. Foot Diameter, be hung up at one End of a very lofty Room, that's 130. or 140. Foot long; and at the other end thereof, (at 100. foot Distance) another Sphere of 4. Foot Diameter be in like manner Suspended, representing the Earth; before which, Lett there be a *Feather* hung by a small Thread as at D. Now if the Sphere at A. (in Immitation of the moons pretended pressure) should in 12. or 13. dayes, according to the moons Semi=Revolution, (or if you will have it, in halfe a Minuite) be moved 5. foot (viz.t) from A. to B. I say can it be Supposed, That that motion, at such a Distance, and in such a wide Room will, by moveing the Air, press the Feather against the Earth at E. And if this cannot be affected; How much less can the Moon at 100000 Miles Distance press the Water, by a Medium that's 6000, times *thinner* than our Air; and therefore, (according to the first Proposition) so many thousand times the more invalid, and *less* liable to be moved by the moon to produce such an imaginary Effect as is pretended.

I could likewise show the impossibility of this Lunar Notion, by Experiments in water; But I pass to the *second* Postulatum or observation.

[p. 72] Wherein we have supposed the Body of the Earth to be inclosed in a Tube, as in Fig: 12.th and the moon at A. pressing or precipitating towards it: In which respect we allow That such a Pressure, (if the moon touched the Sides of ye Tube so as that no Air or Æther could pass out,) might affect the fluid part of the Earth at B. But the Supposition being soe inconsistent with Nature, Reason, and Experience; 'twould be even ridiculous to spend time about it: shall therefore conclude, as we justly may,, That the moon by this pretended pressure, has no more power to *Cause* the Ebbing and Flowing of the *Ocean*; Than the Noise of the Frying=panns, formerly rung by the Old Women, when the moon was Eclipsed; could, as they intended, prevent the moon from being perpetually darkened, or from being removed out of its Orbit.

I shall close this Negative part with only reminding you, That the Dayly difference in the coming of the Tides, on the third day after the ffull and Change, is naturally, but about 30. minuites; Where as on the third day after the first and last Quarters, it is found to 90 or 100 minuts By which it is Evident, That the motion of the Tides is *three* times swifter (or slower,) at one time, than at another. And to confirm that this Phenomena or Observation is soe unequall as herein Asserted, (seeing I finde some *very* fond of the pretended Lunar Regimency) I therefore presume to Challenge, not only such, but also all the Mathematitians and Naturallists in Europe, to confuse the Said: Observation.—As also to prove, That the motion of the moon dos, (especially at its full and Change) correspond with the said Inequallityes; For the small Difference, that is in the moon's dayly motion, (viz.t) from 11. Degrees to 15. cannot in the least be applicable [p. 73] hereunto; Because the said Swift Flowing of the Tides, are in this respect; promiscuous; (to wit) as well when the moon is in her Slowest, as when in her Swiftest, Motion; and therefore has no relation thereunto. But Supposeing it were not Soe, The said Difference of 11. Degrees to 15. is but nearly as $1\frac{1}{3}$. is to 1. which can never correspond with 30. to 90. (being as 1. to 3.) And therefore seeing the pretended *Cause*, dos Not correspond with the Visible *Effect;* we may fully conclude, That neither the Moon's Motion, her pretended pressure, nor her Influence, is the Efficient Cause of the Motion of the Tides. From the whole then, of what has been said, I'll appeall to the imptiall Reader, Whether it be not as reasonable to Allow, That the great man in the moon; dos breath only twice in a Moneth; and then blow upon the Ocean; and thereby cause it to Ebb and Flow; As it is to allow, of the pretended Pressure, or influence before mentioned.——

Sect: **III**

Sect: III.[d]

That the Flowing *of the* ocean *is caused by its being
lifted up from an Horizontall position, once* in 15. dayes
by the pulsing *within the Cavity of the Earth which is
compared to, and confirmed by the* Systole *and* Diastole *of the
heart of an* Animial. *That the Waters being so raised, will
exactly Answear to all the* Phænomena's *found in the flowing
of the* Tides, *proved & Demonstrated by Experiments. And
the places assigned where the Tides are first Moved. By
which the Various* Ebbings & Flowings *may be accounted for.*

HAVING in the precedent Lines, by divers Arguments, *Negatively* proved, that
the moon is not the cause of the Ebbing and Flowing of the Ocean We now come to
The *positive* Cause thereof; Namely to Shew,

"That *the grand Fluxes and Refluxes of the* Ocean, *called* Spring=Tides, *are caused*
"*by the* pulsation, *within the Interiour parts of the Earth. and,*
　　　"Secondly, *That the* intermeditate *Fluxes, are only derived from thence,*
"*as the necessary Consequents thereof.*

Now although we have frequently hinted, at the first of these Propositions; yet it
will more plainly appear, by comparing the said pulsing of the Earth, with that of
other living *Animalls* In which it is to be observed, That as the active or Cordious part
of an Animall, causeth pulsation or flowing of the Blood; in some fifty or Sixty, and in
others, an hundred times, in a Minuite [p. 75] (somewhat according to the magnitude
of their respective Bodyes; as before intimated;) soe the Vast Body of the Earth, by its
Cordious Motion, dos lift up the Waters, and cause a pulsing, or flowing of the Tides,
in a time (naturally) proportionable thereunto; (namely,) once in 15 Dayes; as is before
asserted. Now altho this *pulsing* of the Earth, is near in proportion to that of the
humane Bodyes, as 1296000. is to 1. and seeing her Body if compared with such; dos
farr exceed that proportion; From whence it may be argued, that her pulsing, should
be much Slower, than 'tis found to be; yet inasmuch as all Animalls, (when formed) do
consist, of different matter from the Earth; And one Animall exceeds another, in
Swiftness of pulsing; according to a proportion peculiar to their respective *Natures*, as
well as to their *magnitudes*: As for Instance, if the pulse of a *Sparrow* should be to that of
an *Elephant*, as the disparity of their magnitudes would produce it, the Violence in its
pulsing, would unavoidably convert the matter, whereof the Sparrow was to be
hatched into Ashes; even whilst it was in its Embrio; From whence we may conclude;
that the pulsing of the Earth, once in *15.* dayes, is as natuarall, to the Life or Existance,

of that great Animall; as the other pulsings are to the least Animall whatsoever; tho Such an Arithmeticall propotion cannot be found between them.[96]

Nay 'tis manifest to Sense. That the Motion of the *Tides*, is So diminished between one pulsation, or Spring=tide, and another: That Should the time be suspended [p. 76] for 10. or 15. Dayes Longer, in all probability, The ocean would be quite destitute of any Motion, other than what is accidentally caused by the Windes; and what the consequences of such a Stagnation would be, is not difficult to guess; seeing all Animated Bodies, doe sustain their very being; by, and from, the constant and regular Motion of their fluid parts.

Seeing therefore it is evident, as well by the motion, in the *Waters*, as by what we have said, that there is a naturall, constant, and Proportionable Pulsation, peculiar to the Body of the *Earth*: And forasmuch as we cannot have access, into the interiour parts thereof; to finde out the Means whereby, or the manner how, it is effected,; We will endeavour to show it, by considering and comparing the like Motion, in other Animalls; for we may be well assured: That if the Earth, was to *its* Inhabitants; as [*they*] are, to the *Ticks, Fleas*, and other Animated Creatures, that are generated or produced by them: It would be as easie to know how the Pulsing, is caused in the Earth; as to know how it is agitated in our own Bodyes: Because (as is before observed) it is the same in its kinde, in the Greater, as in the Lesser Animalls. But inasmuch as the Earth is a self=existing, or as independant Body and these are not, We cannot expect to finde, such a Just anologie between them: And therefore it must be obteyned, by a more genearall searching into the Works of Nature; and, (in order thereunto) as we have already shewn how the Earth and its Inhabitants, are Assimuled, in respect of their *Externall* parts; soe we will now.

First *shew Wherein the Earth dos* internally differ *from other Animalls*; and, [p. 77] Secondly, *in what respect it is* assimulated *to them* and,

As to the *first* of these, It is to be observed; That although Nature has So variously formed, the Severall Species of Dependant Animalls; As that some of them, (viz.t) *Crabbs, Lobsters, Oysters*, and almost all Shell=Fish, have their Bones (i.e.) their shells, posited *without* their Bodyes, and their Flesh *within*; and on the contrary, other Animalls have their Bones, posited within, and their Flesh without; Yet I say, notwithstanding these and many other Varieties; all of them are furnished, with Mouths, Entrals, and other Capacities, for Receiving Food into their Bodyes; to supply the Decayes, that are incident thereunto: And not only Soe, But all of them, have their Bloud or fluid part, inclosed and agitated, in Veins, *within* their Bodyes: Where as the Earth is not only destitute of these Capacities, but hath its Bloud or fluid Part *Excluded*, and, (in all probability,) agitated on its Superficies only; So that in these *two* respects, the Earth dos plainly differ from all common Animalls.

Now when I had observed these Varieties in those Animalls and found them thus to differ from the Earth: I thought, If there were any Animalls *Orbicularly* formed; such might, from their uniformity with the Earth, Guide us in this Enquiry, better than those before mentioned; and accordingly, I obteyned the Sight of a Fish, (whilst living, for it liveth Severall Dayes out of Water,) called an *Orbis=Minor*;[97] or Sea=Egg; Which indeed, is soe well assimulated to its first [p. 78] Name, That it was needless to add any other;

But in this Animall, as well as in the former, I observed a Mouth, (tho indeed so uniforme, or Circular, as that it met every way, like the Meridians, at the Poles;) Yet it has no Head, nor any Finns or Claws, to move it self withall: Then I examined the interiour parts; Wherein I observed the Cavity of the Body or Shell, to be Smooth and even; and equally Sphericall with its Superficies; And the greatest part thereof, being filled with Air; the Rest contained the white milky liquor, in which this Oister=like Body, was posited, and Sustained by Severall *Veines* leading thereunto, from the Inside of the Shell; Which was no thicker than a Sixpence, tho it was 4 or 5 Inches *Diameter*

How forasmuch as this Orbicular Creature, as well as those irregular or diversly Shaped, is furnished with a Mouth, and concequently with Bowells, and other Entrails And Seeing tis well known, that the use of all these members, is only to Receive Food whereby to Supply the decays, and Volitians, that are naturely incident, to the Bloud or fluid part of these Animals: And seeing the Earth hath no mouth, (unless you'll have the *Vulcano's* to be Such, which rather sends forth, than receives matter thereinto;) Nor any need to receive any thing within it self, to Supply Such or any other defects; because there can be none, by any dimunition of its fluid parts; For whatsoever is Sublimed in Vapours, dos precipitate on its Body again by Raines; Whereas they continually fly from other Animalls into the open Air, and thence fall to the Ground; And Seeing tis generally granted, That the God of Nature, never [p. 79] made any Thing in Vain; I say from these, and other Like Considerations, it will naturally follow, That the Earth has not any Entraills, inclosed within its Body; as other Animalls generally have; For if it had, it must certainly have a Mouth likewise; Because they Doe inseperably depend upon each other.

Having thus Shown wherein the Earth dos internally differ, from other Animalls; I shall now show, wherein it is assimulated to them; And herein it is to be noted, That although the Earth is wholy destitute of the said parts and Members; Yet it cannot be destitute, of that more nobler part of an Animall, (Viz.t) of an **Heart**, or pulsing faculty which we have ascribed thereunto; for we finde in the Hatching the *Ostrige* before mentioned; not only, That the heart is the first, which is formed, in all Animalls; But also, that such a Cordious motion, is *potentially* in all Matter, even before the Heart, or Entrals are then Educed; as well as continued (by the *heart*.) after the Body is compleatly Animated; And experience teacheth, That the *very* Office thereof, in all Animalls; is principally, to give motion to their fluid parts. Wherefore seeing the Fluidity of the Earth, (i.e.) the Ocean, is Visibly agitated on its Superficies, (whereby its motion is much more conspicuous, than that of our Bloud;) It is but naturall to conclude, That the Waters, are by the same, or Such like corduous power as constantly, and regularly, sett in Motion, *upon* the Earth; as our blood is [by the heart] caused to Circulate *within* our own Bodyes. And therefore, in this respect, the Earth is assimulated to other Animalls; as was before proposed; and accordingly I purpose to Explicate the said Motion. And herein I shall [p. 80] shew, in what manner it is effected, as well *Negatively*, as *Positively*. And,

In the **First**, place we may observe, That if the waters, (as the Bloud in all dependent Animalls,) were agitated, and circulated, by an *Aorta*, and *Vena cava*, (Viz.t) by two, or more, Great Veins, leading [*from*] the two poles of the Earth, (being the places where the waters are first moved as is fully shewn in the Postscript herunto annexed,) [*unto*] two *Ventricles* of an Heart, posited in the Center of the Earth; and by the Systolation and Diastolation thereof, Drawn into and driven from thence, to the said Poles; the Waters would, (besides the prodigious, and inconceivable force, that must be exerted to raise it so unnaturally, for above three thousand miles in a parpendicular height, (to wit) from the Earths Center to its Superficies,) I say it would, if soe driven, be much Warmer, at, and round about the two Poles of the Earth; than it would be, an any other parts of the Ocean: Because the Earths interiour part, must needs, (and is found to) be, much hotter, than the Exteriour part thereof: But inasmuch as this **Phænomena**, is not found or allowed; and seeing the same Motion, may be effected, by a more easie and naturall means, (as shall herein soon be described) we may justly conclude, That the heart, or Active part of the Earth, is not posited in the Centre thereof; notwithstanding it is generally soe seated in all dependent Animalls; And now I proceed to the *positive* part of the Proposition, Namely, to shew *Where*, or in what places of the Earth, its [p. 81] Heart, or Active part is posited; by which the *Tides*, or fluidity of the Earth, is originally and constantly Set in motion.

And in the *first* place, by the way we may learn, from the Forme, and proportion of the aforesaid Sphericall sea Animall; (its Shell being in thickness to its *Diameter*, but as one is to about 250.) That the shell, or ffixed part of the Earth; is not, according to this Computation, above 10. or 12. *Miles* in thickness: Which is soe small in proportion to its Diameter, as that it cannot be Scaled or Represented; unless by the Smallest of the Circular Strokes before inscribed. in Fig: the **VIII**.[98] For which reason I could have no regard to the proportion of the Earths thickness; and as little, to the debth of the Waters; in delineating the same; only I have made the latter, much less than the former; as being most certainly so: For I am well assured, (notwithstanding the Vulgar Notion of the Sea's immensity,) That its Debth in any place, is very inconsiderable; if compared, with the wideness of its *Superficies*: For I finde, between *S⸱Malo* and *Torbay*, and in most other places of the English Channell, that the Depth of the Waters, is but 50 or 60 Fathom;[99] though 'tis there an 100 Miles broad; Which is in proportion to each other But as one foot, is to about 2000: And if compared with the Earths Superficies would be but as 1. is to 360000, and therefore 'tis (as is said) impossible to represent the

said *Debth*, in proportion to the Earths *Superficies*; tho in a much larger figure than that
before mentioned. And from this Instance we may further Learn; That the Debth of the
English Channell, is soe inconsiderable to its Breadth; That [p. 82] There are many
hundred Valies, in *Great Brittain*; which are deeper, in respect of the Hills round about
them; than the Waters are in any part of the said Channell: Insomuch as that if any of
these Hills were pared off and sunk therein; they would stand up, severall Yards above
the Water, tho' Sunk in the deepest places. And as we have thus shewn the disparity of
the debth of the Waters if compared with their Superficies: so we may finde the Lifting
them *up* at the Poles; to be much less in respect of the Earths *Diameter*; For 'tis found by
Experience; That at all Islands in the Open Ocean, which are the properest places for
these Observations; the Water riseth but about 6. or 8. *Foot*: And therefore, doubtless,
at the very places where it is first moved not above 10. or 11. Which is soe little, that it
cannot be described, by the Smallest Line in the said *Diagram*; Though I have therein
made it far otherwise, for explanation's sake. And therefore we may justly conclude,
from the said forme, and proporportion of this *Orbis minor*; if compared with the *Great
Orb* of Earth; That its Crust or Fixed part is not above 10. or 12. Miles in thickness, as
before proposed.

    Secondly, That the Earth, like the said Sphericall or Microcosmick Animall, is
formed internally, with a Smooth Obicular Cavity, and ffilled with Ætheriall matter;
which haply, is soe much thinner than the Sublimations of the *Sun*, as that it is thereby
capacitated, to Float therein;[100] as we see the said Fish dos, in the Waters of the Ocean;
or as a *Buble*, when Blown up, and suspended in the open Air. And also that the said
Æther being thin; will therefore, like other Air, or Æther, admitt of being compressed
and Extended; according as the Motion, of the Earths *Cordious* part, dos every 15. dayes
require; but to return to what we proposed.

    Thirdly That the Active, or Cordius part of the Earth; is of a Flexable or
musculous nature; and withall capacitated to dilate and contract it Selfe, as the Heart of
all other Animalls are; And seeing this Cordious Substance of the Earth, as we have
already Shewn, is not posited in its Center; We may reasonably conclude and Assert,
That it is Scituate, next *under the Ocean*; at, and to the distance, of an hundred Miles,
more or less, round about the *two poles* of the Earth; Being the places were we now
finde, the Watters are first lifted up; as is clearly manifested in the following postscript;
and is accordingly intimated at **AA**. in Fig: **VIII**.[101] where I have Made it full Sea, at
both places, in the said Figure, to shew that tis *so* (and also Low watter) at one and the
same time, at both the Poles of the world. and

    Fourthly That by the *Diastolations*, and *Systolations*; (or Lifting up, and Sinking
downe,) of the said Cordius Substance; the Visible motion of the watters, which we
commonly call the *Flowing* of the *Tides*; is constantly agitated and produced; As has,
and shall be herein further described.

    Lastly that the said Waters of the Ocean being thus raised from their *Equilibrity* Do
thereupon repair, as well towards the *Æquinoctiall*, as to the severall Shoars: In which

interim, (Viz:t) after it has been about 6 ho: and 12 minuits in raising; the said Active part, in the same quantity of time contracteth it self again; and the Waters therewith Subsiding, [p. 84] makes these places gradually become the lowest; whereupon they naturally returns to the said places where they were *first* lifted up; By which returning the watters there, becomes almost as high as when they received their first Elation: on which they repairs again towards the *Æquinoctiall* as aforesaid; and so it continueth, in a diminishing manner to *Ebb* and *Flow* with this once lifting up, for about 15. Dayes or 29. times Successively (viz.t) untill it receiveth a new Elation; and by this means in all probabillity is that admired motion of the watters caused and continued.

Now to confirm what I have before Asserted, I shall prove: *First*, That there are such Active, Passive, and Fixed parts, in all other Animalls, which doe act upon each other, in like manner, as those discribed in the Earth. And *Secondly*, That the before mentioned, Consequentiall or Succeeding Tides, do depend upon every 15.th Dayes pulsation, or lifting up; as before proposed.

And to prove the First, We have an Instance in the Body of an *Eele*, in which it is to be observ'd, That if you Cut out the Heart, and Entraills; and thence Sqeese the Blood; yet the Heart, notwithstanding this seperation, will dilate and contract it self, for the Space of two or three Hours, after it is soe Seperated; and that even though it seemeth, to be as cold as a Stone; as my self with many others have experienced: So that in this Animall, The said three principall parts (viz.t) The *active* Heart, the *passive* Blood, and the *fixed* Body: are manifest, even to Sense. And what is here said of the *Eele*, may be fully understood, of all other Animalls whatsoever; Specially of *Wormes*, and all such, as in like manner, live [p. 85] under the Earth; to whome the Air is of little use, or can have but little or no access. For a conclusion therefore, From what has been said, we may fairly Assert, That 'tis more Reasonable to allow, That the Waters are moved in manner aforesaid; Than it is to allow, the Heart of an Animall, to be the *Cause* of the Circulation of the Bloud in its own Body: For we only feell the Cause, (i.e. the Pulse of the Heart,) and *suppose* the Effect; in the one; Whereas we visibly behold the Effect, (i.e. the Flowing of the Waters,) and *suppose* the Cause; in the other. And that too from what is likewise visible in the Eele before mentioned; All of which being soe plain, 'tis needless to Spend time, for the further Confirmation thereof. Wherefore seeing 'tis evident, that there are, these Active, Passive, and fixed parts, in all Animalls; And inasmuch as the two latter, to wit, the Flowing passive Waters of the Ocean; and the fixed Rocks, Cley, sand, &c: of the Earth; are visible in its Superficies: How can it be denied but that the First, to wit its Active part, is posited, under the Waters to give motion thereunto at the two Poles of the World, as was before proposed: And consequently, That this Globe, whereon we live, is a perfect *Animated Body*, or a Living Animall.

I shall now proceed to the *Second* part of the proposition, Namely, to shew, That the Succeding Tides doe depend upon the first lifting up of the Waters; And herein you may remember, we have asserted the pulsing of the Earth to be only once in about

15. dayes. If so, it may be [p. 86] asked, How it comes to pass, That there are Nine and twenty Ebbings, and as many flowings, in that Space of time. To which I answear, that these intermediate Motions, are naturall, unavoidable, and common, to all fluid (and other) matter, putt, or posited, in *Æquilibrio*, (as the Water in the Ocean is;) whensoever it is therefrom, any wayes moved. and,

To prove this we have verry familiar Examples in Nature (e.g.) In a Plummett of Lead tyed to a String, and hung at large on a pin: or in an equall poized *Beam*, or Scales: For if you Swing the one, or lift up the other, they will either of them, (in a diminishing Manner,) move 20. 40. or 100. times with that *once* moveing only; much like the Motion in the Waters of the Ocean. But that which I found exactly agreeable thereunto, was from an Experim.ᵗ which I tryed with Water it self; being put into two Wooden Shoots or Spouts, one about 22. the other about 4. foot long; Which being respectively at one end, gently moved up, and lett down again, the Water in either of them, did thereupon naturally run to the other end; and being there Stopt, as the tides are, by the Land or by meeting each other in the Ocean, I observed that in its turning, and returning, it would Ebb and flow 15 or 20 times, more or less, with that *once* lifting up: In which it allwayes kept to that naturall and remarkable *Phenomena* in the Tides, (viz.ᵗ) of Raising higher, and running Swifter, at the time and place, of its being first lifted up, and soe gradually decreasing and diminishing, both in height, and Motion, till it quite ceased; which exactly corresponds with what is [p. 87] allwayes observed in the Motion of the great Ocean; as is before asserted, and illustrated, by the *Tyde Tables*, and their explaination: Now the reason of the diminishing difference of its flowing and returning is plain; Because at its *first* lifting up, it is raised farthest, from its Horizontall position; and hath thereby more advantage to hasten thereunto, than when 'tis allmost in Equilibrio: As water will run Swifter down a steep place, than it will down a Stream with a small descent: And as it will then run Swifter, So it will also float any thing faster, and farther, than when the Motion is become more faint and languid. All which doe plainly Indicate that the deep Waters (being posited as aforesaid) are moved up, Only *once* in 15. Dayes; and that their other *intermediate* Motions are only the products or Consequents of the Said Original Motion: or otherwise, they could not decrease and increase, the Difference in their Ebbing & Flowing as before described: For the Dayly difference, in the southing &c: of the Moon, if that governed the waters, is, as it were, allwayed Equall; And the Magnitude, and Light of its Body, unless when Eclipsed, is allwayes the same//

I could now proceed to prove, that the places where the Waters are first, moved, are as we have hinted at the two Poles of the World: And that they do ultimately meet, in, and near the *Equinoctiall* line: as also how they make (in the wide Ocean) a *Floud*, and *Ebb*; at about every 700. Miles Distance; by means whereof there are severall Ebbs, and Flouds, allwayes existing between the Poles, and their Equinoctiall meeting; (it being 5400. miles asunder. 2.ᵈˡʸ That the *Disparity* or great Quantity of water, brought, by the Said Flowings; between *America*, and *Asia* compared with the Parvity so brought, between [p. 88] *Africa* and *Nova Holandia*; is the *Cause* of the Earths Rotation; And that the said Rotation is the *Cause* of the *Trade=Windes*; as also of the *Tornadoes & Monsoons* thereunto belonging. But inasmuch as I have, as is Said already *hinted* at some of these; and *fully discribed* the others, in the *Post Script* hereunto annexed; I shall thereunto refer, and proceed to the *Raising* the four Quarters of the Earth//.

Sect: IV.

Showing *The time When the fishes were Generated; and the
Veins made in the Rocks & stones; also, when the
sea=Fowles were Generated. That the fowr Quarters
of the Earth were raised, by its Pulsing: That
when these, were* raising, *the shells were then
immassed or mingled therein: As soon as the Earth
was* hardned *the Animalls were gradually Generated;
and why it dos not* now *produce such* great *Animalls:
That they were nourished by the* Umbilicus. Why *and*
When, *the* hills *were raised.* How, *and* when, *the
Earths* Rotation *was produced. Why Trees have no*
Females; *& why some Animalls bring forth* many *young
and others but* few. *That the Rivers are caused by
Rains. That the Earth was probably, what we now
call, many years in Forming. That twa's naturally
impossible for the* Deluge *to immass the shells in
the Rocks. And lastly that the Earth, was not
from* Eternity.

It being evident that the Earth has *pulsation* as well as other Amimalls; I shall now shew some further Effects thereof in the forming its own Body. And in order thereunto you may remember; when we at first Spake of Generateing the Earth, we then left it consisting of a [p. 89] smooth, round forme; and of a liquid Substance; with its internall heat and Motion, agitateing within the interiour part thereof. Our buisness in the next place shall be to shew, in a more particular manner, how from this liquid Substance; the four terran parts of the Earth, were raised or increased: But Before we proceed thereupon, must Acquaint, That the Severall Species of Fishes, were Generated [before] and [in] the raising thereof; or otherwise they could not be incorporated therein, and mingled therewith; as we now generally finde them. And in order to evidence the same; It is to be observed, That whilst the Earth existed in this almost fluid State, the Superficies thereof must then, be naturally soft and tender; Whilst that which was nearer the Center, was by the Earths internall Heat, made more Gross and compact; And Forasmuch as the Waters were, as has been said, at the first Creation, Fresh or incipid, and also imprincipled with a *plastick* or Generating Nature; and being (as the form of the Earth was educed,) gradually made more cleer, than when the whole mass was promiscuously mingled together; I say this Water being by these Quallifications, become more fitt for Generation; The Great **Command** of the Almighty, (wherein the said, *Let the Waters bring forth abundantly Fish, Fowles* **&c.**) was *then* (to wit) before the Superficies of the Earth was hardned, or any Hills raised; most admirably and effectually put in Execution; And thereupon, not only the great, and small sorts of Swimming Fishes; as the *Whale, Porpoise Salmon* &c. But also those crawling, and other almost immoveable shell=fishes; as the *Musle, Oyster, Scallop,* Cockle, and the like; were [p. 90] most numerously generated and brought forth; and the Waters thereby most plentifully replenished; as is evident, in respect of the matter that produced them, from Gen: 1. V: 20.21.22.[102] And that this was the *Time,* wherein they were soe produced; is also Evident, from their being to this day visibly immassed in the Rocks, as has been frequently asserted; and will further appear by the following Lines.

The Fishes being thus generated, and, according to their *Species;* either Swimming in the fluid Ocean; (Which then incompassed the whole Earth;) or lying in the Bottome thereof: We shall therein leave them, and proceed to Shew, how the four Terrean parts of the Earth were formed, or raised above the Waters: And herein we must reminde you of the **VI.**[th] *Postulatum* before mentioned, Namely, *That the same Motion, that dos give*

> *Life, and Forme, to any Creature; is continued therein for its Sustentation:* And so, on the Contrary, *That Motion, Which is* continued *for its Sustentation,* is y.ᵉ same *with that, which gave it its* first *Forme, or Modification.*

From whence we may plainly discover; That the Motion which is now continued in the Ocean, is the *same;* with that by which the Earth, was first formed into a Living Body: Which being granted (as it cannot be denyed.) We may thence infer, That as the Waters doe now by pulsation, rowl and overflow the Marine parts of the Earth; soe in like manner, whilst it was forming, it then being smooth & even, and no Land raised to hinder the Same, the Waters did *then* Ebb and Flow, over the whole Surface of the Earth; even over what is now the Land part; as well as the sea. And as in the Generating of other Animalls, their perfection is gradually [p. 91] educed; by the disposing and transposing of the respective matter, whereof their severall parts are formed, by means of Pulsation only: So in Like manner the Body of the Earth in the multiplicity of its fluxings and overflowings; Did transport, or cast up; not only the thin, but also the argilacious and Grosser water, which was in the Bottome of the (now) *marine* parts of her Body, upon what is (now) the *terrean* parts thereof. Makeing then, and thereby, the Beds or Ranges in the Earth, exactly parelel to the Horizon, and to each other, as is described by Fig: 1.st & 8.^{th 103} In which naturall pulsings and rowlings of the said Gross Matter; these Fishes especially that were incapacitated for Locall Motion, (as the *Oysters, Wrincles, Cockles* &c.) and were then, as before premised, Engendred on that part of the said fluid Mass, which is now the terrean part thereof, Were I say, by and with the said soft Matter, covered over, mingled, immassed and incorporated; in such a manner as we have before described, in Fig: the 1 2 and 3:[104] As also Cley, rowled into round formes and hardened into Pebles &c. All which *Fishes, shells, Pebles, sand*, and the like, are now generally found, as well in the Rocks under the salt Ocean, as in and on, the terrean parts of the Earths Superficies; and that in many places of the latter to the Debth of 40. 50. or 100. Foot; as I have frequently observed, and is before asserted and discribed: And that the shells &c. were thus frequently covered over, is evident, because, there are many ranges of them, in the said Debth, one above another, with Rocks and Earth between them; which have noe shells therein, as the Figures doe plainly exhibit: (see Fig: 2.)[105] in which [p. 92] that part marked with **F.F.F.** have no Shels, tho' under and above they are plentifully immassed. Where as, if the forming of the Earth had been otherwise (to wit) by the precipitating of shells and Attomes of Gravell, and Earth together, or such like means; (as some would have it,) The shells, being Lightest, would have been all uppermost; or at Least the Matter, would have been promiscuously mingled; as in Fig: the 7.^{th 106} and Not have been parelel, and evenly seperated, as we now finde them. And this may suffice to Confirme That the *time when*, and *manner how*, the shells were immassed in the Rocks; was whilst the matter was soft, and no hills raised, as has before been frequently mentioned. And the Reason why the said shell=Fish, are found in the Rocks, more plentifull than other Fishes; Is because they were not capacitated to extricate themselves out of the said Muddy matter, as the other Fishes were.[107]

And further to put it out of all doubt, That the mucilaginous Water, was thus transposed or emtyed, out of the Marine parts of the Earth; and cast upon, or Fluxed over the four Quarters thereof; will plainly appear; Not only in that the said *Strata's*, or rather fluxings, are all parellel to each other, in the Bedds or Ranges of the Earth; but also because, that in all cornish Slate or Tyles; as also in almost all sorts of thin Stones; you may see the said *fluxings* of the sea, that were roled over, or cast up in forming them; soe plain, that Nothing but a Confirmation is to be found therein. For as we all know, a small Rush by frequent diping into a fatt, fluid matter, is increased into a Candle: soe in some stones, [p. 93] if but 2. or 3. inches thick, we may finde halfe a Dozen, or halfe a Score, severall fluxings, in the makeing thereof; And some of them may be split, into thin Shivers answerable thereunto: as I have frequently seen and performed. In which it is further to be noted; That as long as the matter thus cast upon the Earth, was *homogeneal*; or was not dryed between each Fluxing; there was no partitions, or Beds made in the great Rocks, nor in some lesser Stones; as is shewn by Fig: 2. But when different Matter, was cast upon the former; or the former matter was

Dryed between the fluxings; it then caused those Horizontall Ranges, or partitions in the said thin Stones, and in the Rocks and Beds of the Earth, which at this day are visible therein; as in Fig: 1. 2. 3. & 4.ᵗʰ is described; and is also evident even by many of the stones in the walls, of the Royall Exchange[108] **London**; and by many thousands of Buildings else where: And from this Fluxing of the Ocean; and at this time, (to wit) before the hills were raised; (for after it could not be;) it was, that the sand, as also the Pebles & other Marine productions, that we frequently finde, as well on the Mountains, as on the more Leavell Parts of the main Land, were cast up and made: For by the Rowling of Lumps of soft Cley, the Pebles were made; and by the Washing of the Earth, the sand was and is still produced, and that this is true, is evident to our *visible* Observation; For on the sea Coast, where any Cley=Hill adjoyneth, in particular about a Furlong to the East [p. 94] of the Passage leading into the Isle of Portlaid,[109] I have seen thousands of Smooth Pebles, that have been made, of irregular Lumps of Cley, tumbled from the adjacent hills into the sea and by its Motion, rowled into smooth Pebles; which in few years have been hardned, into as perfect Stones as those that you would judge to be made a Thousand years before. So that I have frequently with Ease thrust a Stick through some of them; and at the same time with difficulty broke others by reason of their hardness: And that the *sand* was produced as aforesaid is visible to dayly Experience: All Which may serve as a further Confirmation, That the hard Rocks were produced from a soft Substance as aforesaid: And haveing defended to particulars in this; I would also Acquaint, That if any Person is desirous to satisfie his Curiosity, concerning the *Position*, of the Bedds or Ranges of the Earth before mentioned; It is but to walke by the watters side, at Low Water from the said Passage, to Whitenore=Fort, near *Weymouth*;[110] (which is but about a Mile Distance.) or on the Sea coast in the Isle of Portland &c. And he may there see, what I have before asserted, concerning the Position of the Rocks, and Bedds or Strata's of the Earth, and of the shells being immassed therein; as plainly Demonstrated, as any of the Propositions in *Euclids* Elements are, by the Lines and Circles therein contained and Described: And indeed more fully, and far better, that any Words by me devised can express the Same.

The *Marine* part of the Earth being by the said fluxing, and transposing of the *Terraqueous* Matter, (together with the tendancy, intention, and generall designe of Natures Proceess) in some measure emptyed upon the Land; it will follow, That as one part was thus Sunk lower; the other [p. 95] must thereby be made Somewhat higher: And hereupon it was that the Waters began first to be gathered together, and the Dry Land to appear; as we Read Gen: 1. v. 9. 2. Peter 3. 5.[111] Now seeing the Waters were in some measure, thus confined within their bounds; and thereby prevented, from constantly overflowing the Earths whole Surface, as before they had done, we may naturally infer (amongst others) these Three Things (viz:)
First That the Earth, did thereupon begin to Dry and harden into Rocks, and stones, &c. And that in this hardening it Cracked into divers small irregular Joynts or Fissures;

as we see any moist Cley or muddy Earth in hot seasons will doe to this day
Secondly That when the waters, flowed over the whole Surface of the Earth, they must
then, (to wit,) when the Earth was soft; be more gross and Muddy, than when the
Earth was hardned, and the waters gathered together as aforesaid.
And *Thirdly*, That although the Waters after they were thus Collected did not
constantly, overflow the Earth, yet inasmuch as it was then, and is now, at
Spring=Tides; severall feet higher, than at other times, I say, it did, at such Tides,
nevertheless, overflow its whole surface, untill the Earth was gradually raised soe much
higher, as to prevent it there=from:

Now it being evidentt That the Matter of the Earth, must naturally be Crack'd as
abovesaid; and that the Waters at the same time did intermitingly flow over it; It will
as naturally follow, That the waters or other Clear Glutenous and Cristalined matter,
did, and must unavoidably, run into the said Cracks or fissures, and therby fill them up;
Whence, it will also follow, that this must certainly be the [p. 96] Time *when*, and the
Manner *how*, the transparent, and other different coloured Matter,[112] entered into the
said Crackled Fissures; and thereby made those irregular, and Various coloured *Veins*,
in the Rocks of Marble and other Stone, that we now see incorporated therein. And
from thence it was, (namely) From the cleerness of the Matter, That those Veins are
commonly more transparent; than the rest of the Stone wherein they are found; As also
from whence Diamonds, Flints, Marble, and all other *transparent* Stones are, or were,
produced.

And it is further to be noted. That although those Fissures, and other like Cavities,
that were made in the Earth . . . during the time that these Spring=Tides did thus flow
over it Were thereby, and therewith, filled up and incorporated as afore=said. Yet
nevertheless, such of the said Joynts, Fissures, or Cavities, as were Made by the Drying
of the Earth; (for that was the only means whereby they were produced,) after it was
raised so high as to prevent any Fluxing over it at all; I say such Cracks and Fissures
did, and do still remain, open and unfilled: Except those that were Since fill'd up, by
the loose Earth, or by the Factitious earthly Matter, that was, in process of time made
by the Rotting of Vegetables, and Animalls; and has been Since washed or carried
thereinto, by the Rains, and Flouds, running upon the Earth; as is visible to our dayly
Observation.//

Neither is this all the Inferences, that may be Drawn from the premises; For
seeing it is evident, by the precedent Lines, That before the Waters were thus gathered
together, they did constantly, (tho' at first but faint and languidly) [p. 97] Flow over
the whole surface of the Earth; And seeing it is but naturall, to conceive, that whilst
they did so flow, There could be but little or no *inequallity* in their Flowing, to Cause
the Earth (at first) to ponderate, more one way, than another; (For the surface of every
drop of Water, suspended in the open Air, is visibly, at an equall distance from its
Center,) And forasmuch as, especially at every Spring=tide, there was, and is (as is said)

more Water brought from the Poles to the Æquinoctiall, than at other times. **And** seeing by the raising of the Land, the waters were gradually and now are, constantly prevented from coming *equally* round the Globe to the Æquinoctiall; as by consulting the following, or any other Map of the world, and Postscript hereunto annexed, will more plainly appear: I say, What may we hence infer, But that this inequallity in the Flowing and Meetting of the Tides, at the place aforesaid, after the Land was raised; (tho not before,) must naturally bring the Earth, out of its *Æquilibrity*; and thereby cause it to ponderate either from the *East*, or from the *West*; in order to Seek another Center, or place of residence. And being thus Set in Motion: We may reasonably infer, That the **Diurnall Rotation** of the Earth, was from thence first produced; and (by the constant repeating of the same unequall Flowings) is still continued; **And** consequently from thence, (to wit,) From this Gradual Rotation of the Earth, the Dayes and Nights; and even, what we now call, *time* it self, had its Originall and Denonaination. Which I presume may suffice to shew, **how**, and **when**, the irregular, and transparent Veines, and perpendicular Fissures, were made; that we now see remaining **in** and **Between** the Rocks and Stones, that are upon, and under [p. 98] the Earths Superficies: As also **how**, and **when**, the Earths Rotation was first produced.

    But before we pass, from the Waters flowing over the whole Surface of the Earth; to its being fully Dryed, and hardned; it is to be Observed, That the great mountains, and some high Tracks of Land; Must first, (though gradually) be thrust forth, and stand up, (as S.ͭ *Peter* tells us Chap. 3: V. 5.)[113] *in the Waters, and out of the waters*; Whilst all the rest of the Earth, was covered over therewith. And seeing *Nature* is never Idle, we may Well conceive, That when the Earth was in this State, The mucelagenous Water, which remained in Lakes and Ponds, on these higher tracks of Land, did putrify and Generate, all these *Fowles* whose Feet are formed with Fann=likes Claws, to row themselves withall: **And** that the Land=Fowles, whose Claws are Seperated; were Generated with the Animalls next to be mentioned. And that y.ᵉ first, were *then* Generated is plain, Because, tho' they Live on Fish; yet, they could not propagate their *Species* on the waters only, without some Land to rest upon; as dayly experience confirmeth. **Haveing** thus shewn some of the Events; that hapned, *before*, *in* and *upon* the gathering together of the Waters; I shall now proceed to some others, that followed, when the Dry Land more perfectly appeared.

    The four Quarters of the Earth (viz.ͭ) *Europe*, *Asia*, *Africa*, and *America*, being gradually raised so much higher than the Ocean, as that the Waters, (even at Spring=Tides) could no longer overflow the Land; it had thereby the advantage in process of time, by the Earths internall heat and otherwise, to be so hardned, as to become a fit Stage or habitation, for Such Creatures as *God*, in [p. 99] his eternall purpose, had designed it should bring forth; (for before its hardening, as twa's unfitt for them to tread upon, soe it would be in vain to produce them.) Wherefore the Earth being now hardened, and the hills but as yet buding forth, it is but naturall to conceive;

That the misty Vapours and Rains (i.e.) the Sweat of the Earth, that did return and fall
thereupon, whilst it was thus almost level, must, (together with the water and Slimy
matter that was left upon the Earth by the late overflowing of the ocean) unavoidably
remain, in the shallow Lakes and Ponds, that were made by the buding forth of the said
hills and Mountains, in soe many thousand Miles, of almost levell Ground: Which said
slimy Water, and Rainy Vapours, being *then* at rest, and void of Motion, (tho' now the
hills are raised it cannot so remain) it must, I say, by reason of this Stagnation (as dayly
experience teacheth,) corrupt and putrify; by means whereof, it becames a
mucelaginous, Viscous, and Chyle=Like *Menstruum*;[114] Whereupon, (as God had before
Commanded the Earth to bring forth Grass, for the use of Cattle: And the waters to
bring forth fish and fowles,) so he then further Commanded, Saying, *Let the Earth bring
forth Living*[115] *Creatures after their kinde: And it was Soe.* From whence it is evident, that
the Body of the earth; from this, its Viscous or Sweaty *Menstruum* (by virtue of the
Spontanious Nature, infused into Matter, before or upon the said Command;) Did
then, and thereby, naturally engender, and bring forth; First those great Creatures; and
afterwards, as the Hills advanced, those lesser Animalls that do now inhabite the same;
or at least, what is equall thereunto; (namely) their Severall Species. And that this
Spontanious nature, tho' but in a lesser degree, is still in all liquid [p. 100] matter when
Putrifyed; is evident (among other Arguments that may be produced) Because he that
said, *Let there be Light*; and *Let the Earth bring forth*, &c. hath not, in either of these
respects reversed his Commands, Nor have the Effects thereof hitherto ceased; For the
first (viz!) The Light, is Visible, as well on the Surface of the Great Waters, (as has
before been hinted,) as in the Celestiall Luminaries, to this day: And the Latter also,
unless to such as will not see the same: For what is a more evident proff, of this
Spontanious Nature, in all liquid Matter; than that the whole Species, of *Froggs, Tadpols,
Flyes*, and divers Sorts of Insects, are Annually destroyed; and as frequently, generated
or produced again; And although this cannot be denyed;[116] Yet I am not unsensible,
that the manner of their production is Variously assigned: For Some will have it, That
because all great Fowles are Now produced and propagated, by, and from, their
respective Eggs, after copulation; Therefore they infer, That the Eggs which are
produced by these Insects and Lesser Animalls, in one year, are Hatched and brought to
Life, (I suppose by the sun for it cannot be by themselves because they are destroyed,)
in the year ensuing. Now that all Animalls are multiplyed by their Eggs, I readily
grant; For though they were originally, generated without Eggs, or Copulation; yet
being grown to maturity, we finde, that every Vegetable and Animal, has, (as *Moses*
saith) *their seed within themselves*; or other like means, to propagate their *species* by; And
therefore doubtless, in the summer Seasons, these small Animalls, and insects, do
accordingly multiply their kinde: But though this be soe in respect of the great
Animalls, and also, with respect to the lesser whilst the summer heat remaineth,
[p. 101] Yet the Inference will not follow, for the re=production of those that are so

small, as to be destroyed, by their incapacity, to bear the Accidents of the Winter; And
my reason is Because, If the Eggs of those small Insects; or any other Eggs whatsoever,
should be exposed, (as they must then be) to Rain, and Cold; So as the matter therein
contained should become frozen; for so many weeks, or Moneths, as the season
commonly exposeth them; they would thereby, be made wholy unfitt, to propagate
their own Species; Altho we should Suppose them, to remain where they were first
posited; and not carried into the Rivers; and thence to the Ocean; by the Winter
Flouds; as 'tis very probable they are: Which if these, or either of these accidents
should happen; All the severall Species before mentioned, must totally cease; should
they not, (as nature disposeth matter,) be *Spontaniously* generated again; as they, and all
other Animalls originally were. To conclude therefore, untill it can be fully proved,
That the Eggs of tame Fowles, (for others it cannot be so well tryed,) will produce
their *Species*, after they have been so frozen, and exposed as aforesaid: it is but in vain to
deny the Spontanious Nature in Matter; or the plastick power, that has been herein so
frequently ascribed thereunto: For the very Vermine in the Noses of some persons; and
the Various Sorts of Wormes in the Bodyes of others; will plainly Demonstrate the
same, by their being therein produced without Eggs, or otherwise from their like
Species.

I shall not further, at present, enlarge hereupon—only mention one *Observation*
that I finde no notice taken of; and it is this,

[p. 102] That the *Eggs* of all Animals; Do, in a great measure, bear a due
proportion, to the Magnitude of the Bodyes from them pruduced; As by the Eggs of an
Ostrige, *Hen*, *Bird*, &c. (if compared) it may plainly be Discovered and conceived.

From whence, by the way, we may be well assured, That the greater Animalls, (as
has been frequently hinted,) Were the first that were produced after the Creation:[117]
Because as the aforesaid Lakes, or Menstruums on the Earth were at the Creation,
Largest; so the Magnitude of the Creatures thence produced, were, and (according to
the said Observation) must be the *Greatest* likewise: And as we may thus infer in
respect of the Greater; soe we may on the other hand conceive, That the Lesser Species
were generated as the Hills were raised higher: Because the said Lakes, must be thereby
gradually diminished; and also put in motion, (which prevents Generation) by being
changed into running Springs and Rivers; neither of which could be, when the Earth
was almost level; as Reason, and dayly Experience teacheth but,

What I further intend by the said Observation, is to shew, that Persons, (how
curious soever in the Works of Nature.) may Err, in things that are even Visible
therein; and therefore much more, in such as are drawn from *Suppositions* only; as is
that relating to the Insects before mentioned. And my Instance herein Shall be, That
nice Observation, made by the Worthy___[118] wherein he pretends that the *Fœtus*, (and
Consequently that of an Elephant,) is Generated from an Egg, which is no bigger
[p. 103] than a Grain of Wheat or Musterd seed. Now that this is inconsistant with

Nature, as well as contrary to the Said Observation, is evident to Sense: For every Naturallist must acknowledge, That the very Forme of the *Utriculus* together with the two= fould matter or menstruum, thereunto, as nature requireth, constantly coveyed, by the *Vasa præparantia* and **Hypogastrick Veines** &c; is the same in its kinde and proportion; with the Shell, White, and Yelk of any Egg whatsoever: And that the internall heat, of Viviparous Animalls, is also of the same use, in generating any Fœtus, in this internall Egg; as the applycated Heat of the Sun, or Fowl, is to the Hatching any of their externall Eggs before mentioned. And therefore unless another diminutive Egg, or *Hans* in **Kelder**,[119] can be found within the common Shell=Eggs; 'tis very probable, the aforesaid Nicity, ought not in this respect to be regarded: For you see 'tis manifest, that the reall Egg, which produceth the Animall, is the Utriculus, and the said Menstruum, thereinto constantly conveyed; and not the Attome before mentioned: And is also that, by the said constant Supply, which beareth the due proportion to the Body thereby produced, according to the Observation before mentioned. And this I hope may Suffice to Shew, That inasmuch as very curious and inquisitive persons, have I presume erred, in things so visible in Nature; they may much more be deceived in the *re=production* of the Insects &c: before mentioned.

And being now Speaking of these things it may perhaps be Enquired. First, Why Trees, and other Vegetables are produced, but in the *Male=kinde* only;[120] without any Female, to multiply their Species by, as Animalls [p. 104] have: And *Secondly* Why most four Footed Beasts, and other Animalls; have each of them, but only one comon Shell or Egg, (respectively) to multiply their Species in; Wherehas most Fowles have many Eggs, to propagate theirs.

As to the *first* of these I breifly Answear; That these Vegetables, (tho they are generaly of the male kinde,) yet they have no need of any such Female to propagate their Species with: For when their seeds are fallen upon the Earth; The Earth is then of the same use to them; as the shell and heat of the Female, is to the Eggs of an Animall: And the rainy Vapours which fall on the Earth, are also (when putrified and turned into Leffas,)[121] of the Same use as the *Menstruum* contayned in any of the Eggs before mentioned is to the Animals thereby produced. And lastly, the Roots of these Vegetables are likewise of the same use to them, as the Umbilicum is to all Animalls; For thereby the said rainny Vapoures are, not only at first, but also constantly conveyed; for the Nourishment and increase, of their respective Bodyes: Whereas all Animalls, are so Nourished, but only whilst they are in their respective Shells, or *Utriculi*; and being from thence seperated; they are allwayes Receiving & carrying the like facticious Earth, and Water, (tho' more feculent in its kinde) about with them; in their Bowells, and other internall parts; as well for their Sustentation, as for their procreation.

And for the Second Enquiry (viz.t) Why most four footed Creatures, and others, have respectively but *one* com̄on Egg or Utriculus, and Fowles so many; Is because the

Fœtus's of the first are very large; And seeing they must all of them be hatched or brought to Maturity, *within* their Bodyes; it will necessarily follow, That if they should be [p. 105] so numerous, as the Fowles are; it would be naturally impossible for their Bodyes to Conteyn them; And therefore providence has accordingly diminished their Number; and not only So, but hath also proportioned the Lenth, or Shortness, of their Lives, according to the production, of their Severall Species. And hence it is, That those large Creatures, are not only 9. 10. or 18 Moneths, in hatching their young; (if we may so call it) But when produced, many of them Lives to the Age of 30. 40. or 100 years; Whereas many Fowles, do hatch their Young, in a few Weeks, or Dayes; and when brought forth they Live not perhaps, one fifth, or one tenth, of the Age before mentioned; and therefore, as is said, their productions are so much ther fore more quick and Numerous; to the end their *Species* might be thereby preserved. And accordingly it happeneth Annually, with respect to the multitude of Gnatts, and other Small Insects, whose Age, many times is but for a few Moneths, or dayes; and their production doubtless, in as few Howres.

But seeing we have in the Precedent Lines, allowed almost all Creatures, to be multiplyed by their Seeds; it may be objected, that some Vegetables; as the *Vine*, *Elem=tree*, *Fern*,[122] &c have no Seeds, to multiply their Species by. To which I answear, That such as are not there=with accomadated, do commonly, propagate themselves by their Branches or their Roots; as I suppose these before mentioned naturally do; But tho it be So, with respect to these, and such like; Yet I presume, the reason why Some Vegetables, do not in some Contrys, produce seed; is only for want of an hotter Clymett, to bring them to maturity — — For it is evident, that not only divers sorts of Vegetables here; but also the wild or fruitless *Orange=tree*, in *Polonia*; and [p. 106] also some Animalls, as Snails &c: do plentifully multiply their Species by seeding and by Copulation; in extream hot and Dry Sumers; which in hot, and wet ones: (viz.t) when y^e Earth Spontaniously produceth them, they do not So propagate themselves; Nor can we but rarely finde, and such Actions, or productions, by, or amongst them. From whence we may learn, That if the Earth, could have continued, in its primitive fertility; There would not have been, (nor needed) a Male and Female, or any other means, to propagate the Severall Species, that the Earth, at first produced; but the said *Spontanious* nature only, and

Secondly, That if the said Fruitless, or Seedless Orange=tree, or a plant of Fern; or any other such seedless Body; that dos live, or propagate it selfe, *under* the Earth, tho' it be but as it were a lifeless Excresence thereof; I say if any such, were removed into an Hotter Climet; and therein gradually, and naturally cherished (and *not abused by heat*,) with the proper *Leffas*, that its Mother the Earth, to them respectively aforded; it would doubtless, as the (*heat*) or the Sumer approacheth, be so brought to Maturity, as to yeild such seed, as would multiply its own Species: Specially if the said seed should be in like manner sown again, in the same, or alike *Menstruum*, wherein the plant,

Excresence, or *Minerall*, was at first, ripened, and brought to maturity, as before proposed. We might further enlarge hereupon, but I only speak this as an intimation to some persons and return, to the Earth's generating, the Great Animalls before mentioned. And altho, we have already shewn, why the Earth's Productions were more numerous and copious, soon after the Creation; than at this Day; Yet for the further Explication thereof we will Suppose, it may be thus Enquired.

[p. 107] Seeing the Earth did at first *Spontaniously* bring forth such great Creatures; as the *Elephant*, *Camel Horse*, *Sheep*, &c. And seeing she dos now, by such means, bring forth only lesser Animalls; as *Froggs*, *Wormes*, *snailes*, *Flys*, &c. (sometimes) without Copolation: How comes it to pass, or for what reason, is She become more deficient therein, than She was, at her first forming. To this,

I Answear, that there are, as we at first Asserted, only two things in Nature, required for the Generating, or producting all Animated or Vegetated Bodyes; (Namely) *Moisture* or Water, and *Motion* or heat: And as a greater or Lesser Quantity or degree of both these are harmoniously joyned together; the Bodyes thereby produced, are accordingly either greater, or lesser: Inasmuch therefore as the Earth, in the time of its forming, was furnished, as has been shewn, with a greater quantity of Water, (which in putrifying must naturally produce an Heat proportionable thereunto,) than it now affordeth; it will from thence necessarily follow, that she had then, in both respects, greater Advantages conducing to such productions, than it now has: And therefore the want thereof must be the Cause, of the Earth's present Deficiency.

But you will say there's Water enough in the Ocean, and Why is not that, soe productive, as to Generate *Whales*, *Porpices*, *Salmons*, &c. without Spawning or Æquivocall Generation, seeing it did so, when the Superficies of the Earth was all fluid. To this,

I Answer, That it is because, the Nature of ye Waters of the Ocean, is now changed; For when those Fishes were first Generated, it was then, as is shewn at the tenth *Pustulatum*, of an incipid or fresh, indijested nature; [p. 108] But now 'tis throughly Animated; it is so far from generating; as that by its Saline Quality, now acquired (together with its constant Motion,) it even preserveth things from putrefaction; and thereby preventeth Generathion; as dayly experience teacheth; And that Matter is so changed, in its being animated, is evident in that the Miscelanious *Menstruum*, of the Stomack, is first converted into *Chyle*, and thence into the Bloud, which being put in Motion is thereby prevented from putrefaction and Generation, from Whence it is plain, and evident, that the Ocean, being converted and Animated as aforesaid; cannot further any such Generation; other than by the Serous Sublimations or Exhalations thence arising: For nothing but a *Chaos* or Caotick matter; (as was at first Asserted,) can generate a Living Body; And therefore, no Living part of the Body of the Earth, of which the salt Ocean is one, can conduce thereunto; as being now, noe Caotick matter: Whatsoever is dissected, or Seperated, by Exhalation, Sublimation, or otherwise, *from* a Liveing Body; is the proper Subject or matter, for Generation. And hence it is that the Sweat or Vapours of the Earth (i.e.) Rain; as also the Rayes or Vapours of the sun, mingled with those of the Earth; or the Sweat of any other Animall; or any other matter mortifyed, or dissected; as the Earth is by ploughing, or otherwise Seperated *from* a Body; as the Slimy water before mentioned was; (being then in its Caotick State) I say, these, and such like, and no other, are the only proper Subjects or matter, for generating living Bodyes; on the Earth, in the Water, or on the Bodyes of any Animalls whatsoever: And this is manifest, because, That as the abundant Heat and Moisture of the Earth, did first produce greater Animals: so the heat and moisture, of the same Animals; do produce lesser (namely, pedicular) Animalls to this Day. [p. 109] From what has been said, we may plainly perceive, First, That the Great Quantity of Water lying uninterrupted, and thereby putrifying, on the Surface of the Earth, (being also joyned with the Invenial heat thereof,) had then and thereby, the advantage and power, to bring forth those great and divers kindes of Animalls; which now, by reason of the raising of the Hills, and thereby the Speedy runing off of the Said Caotick Water, it can in no wise bring forth; but in a much Smaller Species: And truly, although the *Elphant*, *Rhenoceros*, &c. are very Large Creatures, compared with others; Yet if we compare them, with that Vastly great Animall the Earth, which first produced them; And then compute the magnitude of the *sheep=ticks*; and other pedicular Creatures and internall Vermin, with the Animalls by Which they are produced; We may have cause to Admire, That the Elephant, Camel, Sheep, &c. were not many Thousand times bigger than they are. But the reason thereof is not difficult to determine, if we consider the Heat and Sweat of the Earth; For we finde the Atmosphers of some of those comparitivly, little Creatures, to be 40, 50, and some 100. times greater in proportion, than that of the Earth is.

And in the *Second* place we may learn, That as the Earth did, and dos to this day, produce all Vegetables, but of the male kinde only; so if the *Adamick Menstruum*, which produced, the said Animalls, had been intended to remain always upon the Earth, in the same prolifick State; it would then have produced them, of neither Gender. For as *Nature* never made any Thing in Vain; so it never ordains, *two* effectuall [p. 110] Means, for *one* and the Same End: But the all=knowing God; foreseeing, That the raising the hills, would prevent such productions; Did therefore Cause the Earth, to produce the Severall Species of Animalls; as well in the Female, as in the Masculine Gender; (tho' the latter first,) to the End the like *Menstruums* might, (between them,) be perpetuated; untill the finall Desolution of all Things; or at least, till the Earth dos cease to send forth its Rainy Sweat; for the Production of Vegetables for the use and Subsistance of such Animalls.

Now the manner of Generating and Nourishing those great and lesser Creatures, was at first, doubtless the same, with what we now finde in generating other Animalls by *Æquivocall* generation to this Day: (viz.ᶦ) The abovesaid Heat, being Engendered in the putrifyed *Menstruum*, or liquid matter, lying in Ponds or Lakes on the Earths Supficies, and giving motion thereunto, as is before observed in the hatching Oviparous Creatures; It began to forme or generate the Heart, and therewith the Body and members; and by the *Umbilicall* Veins, (as experience tells us, all oviparous, and other Animalls have;) it administred and conveyed, the aforesaid *Chylous* Water of the Earth, wherein they were generated, for their increase and Nourishment; (for *that* alone is the Mentruum, which produceth the Body; and the Earth but the *Matrix*; as is evident in the Maturating, the Seed of all Vegetables.) untill they were grown to such Maturity, as to break off, the said Umbilical confinement; And being then, and thereupon, capacitated for *Local* motion, they frequently returned to their respective Menstruous Lakes, to Suck (or drink,) thereof: Untill they, (to wit) The *Horse*, *Cow*, *Sheep*, &c. could Sufficiently feed on Grass, which God had before commanded to be brought forth, for their use, or Sustentation: After [p. 111] which the *Lyons*, *Beares*, *Eagles* &c. were brought to maturity and capacitated to pursue, and take their prey; as being First, brought forth for them; and so of the Lesser Species in their order, for there was then, no other Damm to give them Suck; and therefore they must be nourished (as well as produced) in the order, and manner aforesaid: And, (as has been hinted,) we may also from thence, reasonably conclude, That the Beasts of Prey, were the last Creatures, that were produced; (respect being had to the magnitude of their Severall species:) For as the greater Animalls, of what Species Soever, were generated when the Earth was almost level: So were the lesser, proportionably, as the Hills, were raising; only Man excepted, unto whose Wisdome or Authority and for whose Subsistance, the greater, as well as the Lesser of every Species, were, and are Subjected, & made a Prey.

And seeing it is, by the precedent lines, made evident, That the Earth was almost Level when the great Animalls were generated: And that the hills and Mountains, that we now see thereupon, were raised afterwards; (as by the following lines will more fully appear,) we may from hence inferr; That those thinn Stones, or Slates, that are now posited in any hill, how Steep Soever; Did (then, (to wit) when the said Animalls were Generated, and the hills not raised,) all of them lay Levell and parelell to the Horizon; as is discribed in Fig: the 1ˢᵗ and 8ᵗʰ [123] From whence we have this as a *Corollary*, (namely) That if we should be required; to take any such stone or Slate; (though its position whilst in, (or out of) such a Steep hill, should *now* make an Angle with the Horizon, of 20. or 30. Degrees, more or [p. 112] Less.) Yet I say should we be required to take, any such flatt Stone, and put it in the *Same*, position, as it was two or three Dayes, before Adam was Created; it is but to place it parelell to the Horizon, and it will then be posited again; in the same position, as it was, so many thousand years Ago:[124] And this is evident, because the Bedds of Earth and Rocks; were at first, all of them Horizontally Levell, as before Asserted. Haveing, by the way, made this Observation, I pass on, to the raising of the Hills & Mountains Now Visible on the Earths Superficies.

God *Omnipotent* by his previous quallifying of Matter having thus farr, formed the Earth; and replenished it, with Variety of *Vegetables*, and *Animalls*; and by his Omnicience fore=knowing the Effects that would ensue; if the four Quarters of the Earth, should continue in their prestine Smoothness, and fertility; Did therefore cause the internall Animateing heat and Motion, within the Earth, after it was hardned, (as the matter of the Egg must be before the Leggs, Wings, Bones &c. can *bud* forth;) gradually to raise up, and proturbate, these numerous Hills and Mountains, wherewith it is now, almost generally Bestudded, and Indented: Which being fully Effected, the Earth was thereby compleatly Formed; and its *Animation* wholy Accomplished: And hereupon it was, That the Waters were generally gathered together, and the Dry Land appeared: as *Moses* has, most Philosophically, and as elegantly, expressed the same.

Now that those Hills and Mountains may not be esteemed needless, or (as a Worthy Person, in his Theory of the Earth,[125] lately deemed it;) a Deformity, or an accidentall chance; I will therefore presume, tho' it has been already [p. 113] hinted, to give a Reason for the necessity thereof. For if the Hairs of our Heads, are all numbered; and a Sparrow cannot fall to the Ground; without our Heavenly Fathers pleasure; surely much less, can the least of those Hills or mountains, be raised; without the like providentiall appointment. God has not made any thing in Vain; But if we cannot see the *Cause*; 'tis most certainly ours, but not the makers ignorance: One Naturall Reason therefore, Why the Hills were raised, Was because the God of Nature, foreknowing That the heat of the sun, and the internall Heat of the Earth, would constantly cause a Respiration of its humidity, as all other Animalls in sweating do; and that, That Humidity, by not returning to the Ocean, (as our blood to the heart, to be kept in Motion,) would therefore continue, to Exert its plastick faculty; in Generating, and forming such great Creatures, as were at first brought forth; so that in process of time, by the Continuall Generation, and putrefaction, (or rotting) of these great bodys; both Vegetable and Animal; the Land part would have been Increased, and the Sea, by the not returning of the said Exhalations, (or rather Respirations,) so diminished, as to be turned into a fixed Salt; whereupon its motion would totally cease; and an unavoidable destruction thereupon ensue. To prevent which the Hills and Mountains of the Earth; were, as is said, by the internal Energie of its plastick or Self=forming power, gradually thrust forth, and raised up; as naturall as the *Budds*, *Limbs*, and *Knots*, are thrust forth in any liveing Tree, or other Vegetable or Animall whatsoever; That thereby, through the Vallys, the said Exhalations, might finde a more facile, and expedite passage, to return to the Ocean, to be therewith kept in Motion; and the said Fatal Consequences thereby prevented. [p. 114] For doubtless, if the Waters of this Globe or Earth, should Sustain such a Stagnation; or the Earth be any way deprived of its pulsation, and thereby, of its Diurnall Motion; (Seeing they depend upon each other) it would thereupon precipitate towards the *Sun*, and by it be sett on Fire, in like manner as we saw, it happned to the Great Commet that appeared in 1680.[126] Whose brush of misty Raies, or Vapours, (arising from its Calcination,) did, to appearance, when setting, extend to at least, forty degrees in height above the Horizon; And doubtless, whensoever the generall Conflagration mentioned in Malachi 4:[127] shall be accomplished; (If it be not miraculously effected,) it must be, by such means, as we have now described.[128]

And that this *encreasing* the Land, and *diminishing* the Sea; by the means aforesaid, is no groundless Supposition, will appear even to a visible Demonstration (namely) because the Earths Surface, is generally to the debth of 2. or 3. Foot more or Less, covered over, with a kinde of Black or other coloured Garden=like Mould, though it be Rocks, Cley, and other different matter immediately under it: Whereas the four terreane parts of the Earth were raised, by transposing or emtying its Marine Parts thereupon; And therefore it cannot be supposed, nor is it found by experience, That the Bottome of the sea did, or dos, consist of, or could afford, such sort of Mould to cover its Superficies withall: And therefore it must be otherwise acquired, which we conceive was in this wise, (vizt) The Earth being at first covered over, in some parts, with such matter as produced *Cley*; which in drying Crackled into small peices, and were afterwards hardned into Stones; In other parts, with such matter as is now become sand, and Gravell; a Third Flint, Pebles, &c. [p. 115] I say the Earth being at first thus formed, and, as has been shewn, also covered over with divers Lakes and ponds of Water, which then plentifully produced divers sorts of large Vegetables and Animalls; who liveing but for a Time, and so decaying; were afterwards rotted and turned into such Mould or Earth as is before decribed: and by its soe plentifully doeing at first, and still continuing, (though in a much lesser degree) in generating and decaying; for so many thousand years;[129] the said mould is thereby increased, to the debth of 2. or 3. foot as we now behold it. And that it was by this means Increased, is further evident, not only because to this day, the more fertile the Ground, the thicker the mould: But also, in that the Grass, Corne, Wood, Bones, shells, &c; (all which, is but only Water fixed.) that is produced, in one year, by the said Serous Vapours of the Earth and exhalations of the Ocean, (of which who knows how many Ton is required to make one Ton of Timber;) is not, or naturally cannot, in the *same* year be rotted and turned again into its originall ffluidity; without which, it cannott return again to the Ocean, from whence it dos constantly proceed: And therefore, Whatsoever was or is, Anually left remaining; either of Animalls, or Vegetables; did, and do, still encrease the Debth of the Earths surface in manner aforesaid. Whence we may well conceive, that the Cause, why the Hills were raised, was to hasten the returning of the Water; to the End it might be prevented from being embodyed, into such large Creatures, as would have encreased the Earth and diminished the Ocean, even to such [p. 116] a Degree, as would have exposed it, and us, to the destructive consequences before mentioned. and.

Haveing now hinted, that the waters do originally proceed from the sea, it is to be observed, That many have affirmed,[130] That the Springs and Rivers are produced, by means of the Waters passing from the Ocean, through the Earth; And Not by the

Clouds, and Rains, falling thereupon. Which Doctrine I conceive to be inconsistent with Reason and Experience; For, as the Wise man affirmeth and our Experience teacheth; *all the Rivers run into the Sea*; Therefore their Springs & Fountains must needs be *higher* than the Ocean is, whereinto the Rivers disgorge themselves; or otherwise they could not desend and fall thereinto: And inasmuch as it is manifest, that water cannot raise its self, above an horizontall position; and the spring head, must, as is said, be *higher* than such a position; the sea can by no naturall means, be the Author thereof, otherwise then by affording its Respirations, or Raine, to fall upon the Earth as aforesaid. Nay, I would desire the wisest *Pholosopher*, or Naturallist; to Shew a Reason, why the Waters of the Sea may not more easily run to the head of the spring, by the open mouth of the River; than to press thro the Pores of the Eearth; and unnaturally raise its self, above its Horizontall Levell; that thereby it might be capacitated, as is falsely pretended, to return again to the Ocean; I could further confute this Doctrine but a Word to the Wise is sufficient,

It being in the foregoeing lines made evident that the Earth was hardened into stones &c. before the Hills were raised, and the Cause of their raising being likewise declared. [p. 117] We now come to Speake of the *Time*, that (according to probabillity may naturally, be required or spent in forming the Earth; And herein it is to be observed, That an *Animall* (the largest whereof, is but as an Attome, if compared with the Earth) is not, out of a lifeless liquid matter, converted into a Liveing Body, in an Instant; but in some measure, a Time proportionable to its Magnitude is required, for its Generation; as some of lesser kinde in one Month; some of greater in eleven, and the Elephant in eighteen; so neither must we imagine, That this vast body of the Earth, consisting of 21600. Miles in Circumference, could, according to the course of Nature (which is the only means that we doe herein pursue, and not the power of the Creator,) be thus converted from a soft fluid Substance into hard Rocks, Cley, sands, &c. in a *Moment*; but that some naturall proportion of Time, according to its Magnitude, might be thereunto required.[131] For if soe small a Creature as an Horse, Cow, &c. dos, as is said, require almost one year for its forming and hardening into Bones &c: surely the Earth which is soe many Thousand times greater, might probably require what is now equall to many Years for its forming: And therefore the Words of Moses, when he speake of the *six* dayes Labour, must (as I humbly conceive) be understood as when parts, passions, and Members are ascribed, to the Creator of all Things; (viz:ᵗ) but only to Sute or adapt Things, to the weak capacity of humane understanding: For seeing it is evident from the Words of sacred Scripture, That a *Thousand years* with the Lord,[132] are but as one Day, [p. 118] why may not the *first* four Dayes, (to wit) untill the Sun and Moon were made; (see Gene: 1. V. 16.)[133] or till the dry Land appeared, and the Earth attained its Diurnall Rotation; or at least whilest the Light was produced by the spirit of God moveing upon the face of the Waters: I say, why may not one such day, be equall to many years: For seeing, as has and will be shewn, The Rotation of the Earth, is caused by the Disparity in the Flowing of the Ocean, And inasmuch, as whilset the Earth was *all* Fluid, (to wit) till the four Quarters were raised, there could be no such inequallity, or Disparity in its Flowing; And therefore no Rotation; and Consequently no measure of time: (Because *that*, is measured only by the Earths Rotation,) From all which, I say it is very probable, That the first Three Dayes, might at least be many hundred times Longer than now they are; Therefore &c: And this *slow*, or *want of Motion* in the Earth, will in some measure appear, from the Motion mention in the said Egg; which at first is so languid, as to be Scarce discernable: Nor is there a perfect Locall Motion attained; untill the Body is as it were, Maturely Generated: But that which makes it further evident is from the great *Oyster shells*, and others, That are found immassed in the Rocks; which though they doe naturally require severall years to attaine such largeness;[134] yet they must be fully grown whilest the Earth was generating; or otherwise, they could not be incorporated therewith, & immassed in the Rocks as aforesaid; And, (which is more,) The Fishes that were in the Shells, at the forming of the Earth, must be quite Rotted, before the thin, *cleyish*, *Stoney*, and *Mettallick Matter* could enter into the shells to fill up the Cavity wherein the Fish formerly lived; (thereby makeing the [p. 119] Forme of the said Fishes, of perfect stone &c:) as we now plentifully finde them: And how Long Time such rotting (or as it were annihilating) would naturally require, when soe covered from the Air, Who is he that can determine? Nevertheless, seeing the truth of what I have, (in Generall) advanced, dos not depend upon this particular; I leave the reader to judge thereof as he pleaseth: Now if the Earth, was, as is very probable, many hundred years (or rather what was equall thereunto) before it was fully formed: And inasmuch as all great Animalls were generated before the Hills were raised, (tho' not before the four Quarters of the Earth, were raised some what higher than the Ocean, as before is shewn;) It will necessarily follow, That whilest the hills & Mountains were *gradually* buding forth, (for Nature abhorreth Violence) there must be in that long space of time, a very plentifull Increase of Land Animalls and Vegetables, (as well by Copulation and seed, as otherwise;) as we finde by the Rocks &c. there were of fish=animalls before the four Quarters of the Earth were raised; Which being granted; we may easily learn, that the *Bones* and *Trees*, that are yet to be found buried and mingled with the Earth, (tho' not immassed therein; because they were not brought forth *till after* the Earth was hardened,) were such bones, and Trees, as happened to grow or stand upon the Edges of the Hills, at the Time when they were almost or fully lifted up: Which Hills being raised to such an height; as some of them to make an Angle with the Horizon of 50. or 60. degrees in Altitude; I say for such Hills as were soe very high raised, some part of them, when come up to such an height must unavoidably tumble down; and haveing then Trees, or Animalls, standing thereupon, (tho the Trees stood perpendicular before the falling of the Hills) must of necessity fall down with them, and be buried; with whatsoever Rocks or Rubbish, did accompany the same; as the Trees &c: standing on the hill at **A.C.** Fig: 3ᵈ¹³⁵ do plainly intimate. And that the Rocks, and Earth, did thus tumble down, you may see Visibly manifested, (as you Ride [p. 120] on the Rodes) in almost all Steep hills whatsoever, which I desire may be accordingly Noted: For at that End Which the Ranges or Strata's cuts the surface, (as at A.C.) the Rocks & Rubbish, that have so fallen, do in many places, yet remain in that fallen, or irregular position, to this day, though the strata's in the remaining part of the Hills are still exactly parellel to each other, as before described.

And for further Confirmation, That the Earth was not replenished with the larger sort of Animalls and Vegetables, untill *after* the Matter was hardened and fit to be raised into hills, as has before been shewn, is evident from this *Phænomena* in Nature; namely, That notwithstanding there are, as is said, such plentifull Numbers of shells, immassed in the midst of those vast Rocks; yet we can rarely finde, any such Trees or Animalls incorporated in them in such manner as the Fishes are; Whence 'tis plain, that these Trees and Land Animalls must be generated after the Rocks were hardened: And Consequently were tumbled down by the raising of the Hills as aforesaid; or otherwise they could not be found, only mingled with the Rubbish, that fell from thence, and not incorporated in the Rocks as the Fish shells are. and,

As for the Manner how the said Land Animalls, and Vegetables were *preserved*; It must either be, from their being covered up in the said Rubish, and soe allwayes kept in one State of Moisture; or else, their sap being exhausted, the then petryfying humidity of the Earth, filling their pares, preserved them, by converting them into a Stoney Substance: And for the Confirmation hereof, we have an Instance of the first, Way of preserving them, at the Bottome of a very steep Hill near *Shaston in Dorsetshire*,¹³⁶ where I lately saw severall Trees that were digged out of the ground, adjoyning to the Foot of such a steep Hill, (one whereof was 20. or 30. Foot long, and about a Foot Diameter, which doubtless was [p. 121] tumbled down and preserved, by being covered over some depth under the Earth, as was before proposed; And of the Second sort (*namely*) *Petrafied Wood*,¹³⁷ may be seen in severall places in this Kingdome, in particular amongst the Curiosities of Grasham Collage; And doubtless in this manner especially by the first were the Bones of Animalls, (being of equall or greater duration) likewise buryed and preserved.

Having thus given you my Thoughts concerning the forming of the Earth, and how the Shell=Fishes and other Animalls, and Vegetables, were produced therefrom, and involued and preserved therein, I now come to Speake a word or two concerning the *DELUGE*¹³⁸ of Water that was brought upon the Earth, in the Dayes of *Noah*; And herein you may observe, that I have wholly excluded, that Flood, from haveing any share, in bringing the said shells into, or upon the Earth; and that notwithstanding the Opinion of many learned, and otherwise ingenuous persons to the contrary; who I perceive finding the shells in and upon the Earth; and haveing been told, either by *Philosophers* or common Fame; that the Said Deluge brought them thereupon; have therefore taken the same for Truth: And to Reconcile their observations to the said Report, or to their own Opinions; have Rack'd their Braines, and as it were unhinged Nature; to make the said Flood the naturall Cause thereof: But if because there are shells found in, and upon the Earth; and only because common report tells us, that the Flood brought them thereupon; when the *Sacred* Writings, are silent therein; and *Nature*, and *Reason*, will not admitt thereof: I say if we must believe it, only because it is soe reported; we may as well be imposed upon to beleive the fabulous Stories of a Generation of *Faries*, and their circular Danceings: For as we have shells found in and upon the Earth; soe we have likewise green Circles, frequently found in the [p. 122] Grass, which common Report, tells us were made by the Foot Steps of that pretended Generation. Whereas if we examine into the Cause thereof, we shall finde (If it were but only by cutting up a Turf, in the said Circle) that their originiall, is not from the footsteps of those or any other Creatures. And not only from thence but inasmuch as the Cause thereof has been otherwise fully accounted for, by D.[r] *Plott*[139] and some others, we may therefore reasonably conclude, That the very being and pretended footsteeps of such a Generation, was at first, only a Poeticall or Fabulous Invention; and perhaps designedly continued, by the popish Priests, before the Reformation (for since tis not generally believed, tho' the Circles are as common as before;) to amuse the Vulgar; That thereby, they might carry on their Cheats with the greater security: fforasmuch therefore as this common saying, was Certainly Entertained, through Ignorance, and too much credulity; soe the opinion of the Antient, and modern Philosophers, concerning the *time* when the shells were incorporated, did in like manner proceed from a like credulity; or at least, from the want of a due Examination into the Reason of their being posited, as before Described. But tho' I disallow the Flood to bring the shells &c. into the Rocks, and the very being of such a people; Yet I would not have it thence inferr'd, that I disbelieve the being of such a Deluge, as might Destroy the Inhabitants of the Earth; it being not my Intention to disallow thereof, or to intermedle therein: What I chiefly contend for is this, (viz[t]) That the waters of the Deluge, if they were what we now call waters, could not naturally, (for we pursue no other means) at that Time, nor at any other, dissolve the Rocks and Mountains of the Earth, and mingle and immass the shells thereinto, as we now finde them; notwithstanding it is so generally allowed by others, and that [p. 123] for these further Reasons following, amongst others that may be offered.

First, **Because** if so, The water must then, be much stronger, and more Corrosive in its nature, than *Aqua fortis* or any other Liquor, of what nature or kinde soever,; For otherwise it could not have such an Effect, as to dissolve the Rocks & Mountains as is Vulgarly Supposed.[140]

Secondly, If the waters at the Deluge, were of such a nature, Then the whole *Species* of Fishes, how large soever, must unavoidably have been thereby totally destroyed: And forasmuch as no *Whales*, nor other Fishes, were preserved in the *Ark*; There must have been, If not a new Creation of them, at least a new plastick Nature, or forming power, infused into the waters; or otherwise it could not have produced the Variety of Fishes that we now Finde living in the Ocean. and

Thirdly If the waters did then so dissolve the Rocks and Mountains; there could have been no Mountain standing at the Deluge, for the Ark to rest upon; as we finde there was, even whilest the waters were upon the Earth: see Gene: 8th. V. 4.[141] and

Fourthly, It may be thus argued;

1st Either it *was*, in the nature of Water, ⎫

  or   ⎬ to dissolve the Rocks and mountains of the Earth.

2dly It was *not* in the nature of Water, ⎭

If the *first*, then tis no less than a Miracle, that they were not dissolved a Thousand years before the Floud: or else some Thousand years since: As also that they are not *so* dissolved, at this very Instant, and,

If the *second*, Then tis as great a *Miracle*, That they should (as is pretended) be thereby Dissolved, at the time of the [p. 124] Deluge. But that they were not Dissolved, *before* the Floud; Nor at any Time *Since*; Is not, nor cannot be denyed. And That they were not Dissolved, *in* the Floud is Evident from the third Proposition beforegoeing.

Wherefore the said Rocks, and Mountains; have not, either Naturally or Miracuously, been Dissolved, ever since they were first hardened *in*, *at*, and *after* the Creation.[142] And therefore the shells could not be admitted thereinto, at the Deluge, or at any other time, But only *before* the Earth was hardened, as is before described and Asserted: I shall not further enlarge hereupon, other than Appeal to the impartiall Reader, Whether from what is contained in the foregoeing Lines, it be not more Reasonable, and Naturall, to conclude, That the Earth was Generated from an Aquous fluid matter, and thence hardened, into Rocks, sand, Cley, &c; and the mountaines raised after it was soe hardened: And That the shells were immassed or incorporated in the Rocks, *at the Geneating* of the Earth, when the matter was soft and tender; and no Mountains raised; as is before Asserted: Than it is to allow it to be done after the Earth was hardened (to wit) at the time of the *Deluge*; which must leade us into so many improbable Consequences, and unavoidable Absurdities.

Having thus shewn, how the Earth was from a fluid matter formed, and converted into a *Living Animall*; and how the shells were brought thereupon, and immassed therein; as also hinted at the naturall Cause, of the Ebbing and Flowing of the ocean, *How* and *when*, the Mountaines were raised, &c. I shall at present conclude, with this one *Theologicall* Use, drawn from the premises (viz.^{r})

There are, as in times past, many who say, (at least,) in their Hearts, *There is no God*; Psa: 14 V. 1.,[143] That the Earth, was not Created, nor had any beginning; but was ever the same as it now is, and will soe [p. 125] continue, even to all Eternity.[144]

Now altho what is before delivered is sufficient to prove, That the Earth had a beginning, and is Therefore not eternall; yet for the further Confirmation thereof, I will endeavor to prove the same, by putting my Argument, into this *Syllogisme*.

Whatsoever was, or is, from *Eternity*, or had no beginning; was or is, *eternally the same*, without the least Variableness, or changing whatsoever; either in *Nature, forme,* or *substance*.

But the Earth was once, a *smooth, soft fluid matter*; and is now converted or changed; into *hard Rocks, stones, sand, Mountains, Hills, &c.*

Ergo; the Earth was not from *Eternity*.

The Major, being undeniable, and the *Minor*, demonstrable; even to *visibility*, (from the fish=shells &c. being immassed in the Rocks as aforesaid,) *The Conclusion* must be unavoidably granted.

Wherefore, to that Eternall Existence, that Created, commanded, or brought forth *Matter*, out of *nothing*; and also infused thereinto its *plastick*, or self=forming power; and thereby Generated or made, not only this *Globe* of Earth, but even the whole *Universe*, and all Things therein contained; be Glory, Honour, and Adoration, for ever and ever. *Amen.//*

Finis:

(1) [126]

A Postscript

Shewing the two places where the Tydes are originaly
moved, and where they meet; as also the mañer of their
Flowing.

'Tis almost ten years Since I composed what is generaly contained in the foregoing Treatise;[145] and although I was well assured from observations and reason, before I entred thereupon, y^t ye waters in ye Ocean were made to Ebb & Flow, by being lifted up, in some certain place or places thereof, as is before Asserted, & not by any *Lunar* Attraction or Pressure Whatsoever: yet when I came to treat of the Said places [*where*] they might be So moved; I then found, That that Discovery could not be attained, unless a Generall Account of the Setting of the Tides, in the Severall parts of the Ocean, could be procured; wch I then dispaired of; & therefore proceeded to ye other Sections; Leaving ye sd place or places, where ye waters were moved undetermined, and ye whole, in that respect, Imperfect: Tho' since I have Conformed it, as Occasion required, to ye following Discovery, But notwithstanding I then, as is said, thought it Unattainable; yet I frequently wished & Endeavoured, for Such a *Register*: And ye rather, because during all this Intervale I could never see any *Phænomena* in ye Tides; Nor any Character imprinted by nature in ye earth; but w^t Ecco'd to & Confirmed, w^t I had conceived thereof, & is before delivered: Now haveing, I say, thus Endeavoured, I accordingly, Some few months Since, met wth a large Folio, entitled ye Lightning Collumne or Sea Mirour; Printed [p. 127] at Amsterdam Anno Dom̄: 1701.[146] In which I found a full account of the Setting of the Tides, for almost all the Sea Coasts, in the Westren or Atlantick ocean and North Sea, plainly entreed and Recorded: Which Volumn you will find to be industriously compiled, from the Observations and Experince of Divers Mariners; (especially of the Dutch,) as they had in many Ages, in their Severall Voyages entred the Same. But when I came to read these Various Accounts: I still found a great Difficulty in this Enquiry; For although they were plain and perfect; Yet were they so numerous and promiscuously placed, as that it was imposible to forme or concive, any Canonical or Regular Idea, of their *General* Setting, by the bare reading Thereof: Because as fast as we Read of their Setting on one Coast, those of the others, would be either forgotten; or at Least become confused. Whereupon after some thought and Reasonings on these Difficulties, I conceived there might be some Device to Supply the Said Defects; and in Order thereunto, I prepared a Large Mapp of the World,[147] of which this annexed is a Compendium, whose By=Section I made under the Meridan of Natal in Africa; and placed the Continents, in respect of Longitude, according to Mr *Mol's*[148] and others, late Discoveries; as conceiving Such a By=Section most proper for Exhibiting the *Marine* parts of the Earth: And when I had thus done, I look the said Volumn, and what was incerted therein,

Relating to the Course or Setting of the Tides, I forthwith Delineated, and entred into the said Mapp, by only placing the Characters of small Arrows, with their heads pointing which way, and on what point of the Compass the Flouds do sett, on the Several Coasts, as fast as I read them: Whereupon I had by means of the said Arrows, at one View, or cast of the Eye [p. 128] whatsoever was contained in the whole Volumn, relating as aforesaid, plainly represented and Described. And haveing thus Succesfully obtained, So Compendious a contraction, and made such an instantanious Representation, of what was So Volumniously Treated of, and So many Ages in Collecting; I resolved to prosecute the same in other parts of the world: And accordingly I procured other Lightning Sea Columns, and Severall other Volumns of Navigation, as M.ͬ Sellers,[149] and other, English Pilots: M.ͬ Thornton's[150] Oriental Navigation in Folio, M.ͬ Dampiers,[151] M.ͬ Roger's[152]—and other Voyages round the World; and after I had in Like manner entred the Characters, according to what was recorded in those Volumns: I Visibly perceived; by the pointing of the Said Arrows, that the waters in the North Hemisphere; do generaly Set or Flow towards the South: And on the other hand, those of the South Hemisphere; do in like manner, Generally flow towards the north: From which Visible and undeniable Phænomina (with a kinde of admiring Speculation,) I plainly discovered, *That the watters of the Ocean, are orginally moved at the Two Poles of the world: And do ultimately meet, in or near the Æquinoctiall Line.*[153] Now that this is the genuine and Natural motion thereof, is, in Some measure, confirmed, from the very shape in which the Earth was originally formed. (Viz.ͭ) in that the two grand Continents of Land, and the two great Oceans of water, do all of them, range a long with the meridians, from North to South; which was doubtless, by nature, So designed, on purpose to admitt, and further the Said Motion: For if they Should have trended, from East to West; it would have interupted, or wholy prevented the Same. And what dos further Confirm this Doctrine, is, In that the many great mountains, or Island= like sholes of Ice, that are annually found Swiming in the North part of the Atlantick Ocean, are always observed to Soar along, and make their way good, as 'tis called, from the North to [p. 129] the Southward; even till the heat of the Sun has totally dissolved the same; as the mariners, treading to those parts, do generally testifie: And doubtless, the like Sholes would be found, coming from the South Pole, towards the North, if our Navigation did lead us there, to make the like observations. Another Instance which confirms the Said Doctrine is, in that the Tydes, do generally run much Swifter, and rise much higher, in these and the more Nothern parts of the world, than they do between the Tropicks, and near y.ᵉ Æquinoctial Line:[154] Now that this dos prove the Said Proposition is evidenced by the Experiment of the water in the Sute; mentioned in Sect: **III**, whereby it appears, That water will naturaly Run Swifter and rise Higher, at, and near the place where it is first Sett in Motion, than it will at a greater distance: From whence it is evident, That as the knowledge of the place where it is first moved, dos now show us the *Cause* of the Said Phænomena, to wit, of its rising highest in the Nothern parts; So the said Phænomena, dos as plainly demonstrate, That that must be ye place where the Tydes are first moved in the Nothern (and, doubtless, in the Southern) Hemispheres: By all which it is manifest that the motion of the water is such as is herein before and after described.

But if notwithstanding what has been Said, any person (through their affection to the immaginary and medly Influence of the Sun and Moon,) shall object, That I might be partial, and not Sincere, in those deliniations; and consequently, disbelieve the truth of this discovery; my Answer to such is, That I refer to the Said Vollumns for my Iustification; with this confidence, as to challenge all the Philosophers & Mathematicians in Europe, to Confute the Said Doctrine; either by the Vollumns before mentioned, or by any other, (if such there be,) of like nature. And as for the Sincerity of the Mariners, that recorded those observations, it being done in So many Ages; and by Such Variety of persons, it would be absur'd to Suggest, that they should be fraudulent therein; as not knowing [p. 130] What they had entred, would ever be used, to produce or warrant Such a Discovery: And therefore, if it should be Suppos'd, that I have been Misled in any particular, it must proceed from their taking a Trade=wind=Current, or the like, for a Natural Tide; & not from the Partiality or designe of the Said Authors, or my Selfe. And as for the Objection that may arise, on account of the waters Setting with the Trade=Winds, Monsoons, &c: I shall briefly Answere, That that Motion, is only on the Surface of the Ocean; and that under its Surface, the Tydes have their natural Course, as before described: and accordingly M^r. Dampier[155] has told us, That he found, under the Surface, a different motion therein, (by his observations) when under the Clymet, where the said winds prevail. Inasmuch therefore, as the Vollumns before mentioned are impartially compiled, and the Delianiations in the Mapps, accordingly Drawn; and consequently the aforesaid Proposition, thence arising, fully confirmed: I shall now pass on to the further Explication thereof. And as for that part of the Mapps which relates to the general Setting of the Tides; it is So visible, That 'tis need less to Say more, than only acquaint; That as the Arrows heads, do shew which way the Water flows, 'tis but naturall to conceive, That it generaly Ebbs, on the opposite point of the Compas: And where you See them pointing one against another, as on the East of Ireland, and on the SouthEast of England; there the Tides do accordingly meet each other. And as we know, there is no concussion at their meeting, in these, and other like places; So wee may be assured that there is none, at their returnings or meetings under the Equinoctial, or at ye Respective Poles of the world; and consequently the waters are there as Navigable, as in other places of [p. 131] ye Ocean: And this I hope may Suffice in brief, for the Original, Course, and Meeting of the Great Waters; as also to Shew by what means I obtained the Discovery thereof: (to wit) From what is Recorded by Mariners, in their Several Observations as aforesaid: And Seeing this Method wherein I proceeded; is the Same by which M^r. Halley, in his Late Mapps,[156] has delineated the Variation of the

Compass, and the Trade Winds; I See no Reason, why this Should not be equally
accepted and approved. I Shall now proceed to Some other Observations, that
Succeeded as Corollaries form the frist Discovery; For as in Algebraical and other like
operations after the Aquation is discovered and the Solution brought to Light; there
frequently ariseth, Such Consectarys and Aphorisms, as before were not expected; and
Sometimes Such as could not be preconceived: So when I had discovered the places of
the Tides originall, and meeting: I likewise found the *Cause* of the manner of their
Flowing; which could not So well be known, or conceived, before this Discouery was
made: As will plainly appear by the following Lines.

And in the First place it is to be observed, that almost on all the Sea Coasts, from
Ushant to Cape *S.ᵗ Vincent*, and thence even to the *Canary* Isles; (as you will find by the
Said Register,) a S:W: and a N: East moon, makes, (as tis called,) full Sea, for above a
thousand Miles together: And not only So, but tis always full Sea, all along these
Coasts, Severall hours (*before*) tis So, further up in the Rivers thereunto belonging;
whereas in the English *Channell*, as also in the *Mediterranean Sea* and other like places;
where the waters runns (by) the Land, tis in both these respects quite otherwise; for 'tis
not only ten or Twelve hours difference, in the time of full Sea, in the space of but 3.
or 400. miles, in the Said Channell: But 'tis also full Sea in the Rivers & Ports thereof,
Severall hours, [*before*] tis So in the Offin or channell it selfe: all which will plainly
appear, by the hours shewing the time of full Sea in figure **III** Plate 3.[157] or if compared
with the Register before mentioned.

Now the Reason of this great Difference in the flowing of the Tides, will easily be
accounted for by this Doctrine, and by the Delineations in the Said Mapp; For tis there
made evident even to inspection, That the Flouds on the Said Coasts of Portugall &c;
do generally Sett, or devolve, directly upon and against the Land; as it were at One
and the Same Instant; For tis not above a point or two Difference, in the 1000 Miles
before Mentioned; Whereas on our Coasts, it runns along [*by*] the Land; and is
therefore 10. or 12. hours (i.e.) 14. or 16. Points, before it can run up, from one end of
the Channel, to the other; as before Observed. Whence we have this as an Aphorisme,
That wheresoever you See the Waters Setting [upon or against] any Coast; and
running [*by*] another; you may certainly Conclude, That the manner, and Difference
of time in their Flowing; is such as before described. And this will hold true, even in
the midst of the Oceans, as well as else where. The Second *Phænomena* to be Observed,
is the Violent Course or Setting of the waters, in the *Gulph of Florida*; Where it runns
So Strong to the Northward, for Some hundred miles together, that ther's no Sailing
against it, though favoured by the Strongest Winds conducing thereunto: Which has
Seemed to Some a *mistery in Nature*; But by this Hypothesis 'twill also be easily Solved,
& appear to be only naturall: In order to which it is to be Noted, that That part of the
Ocean being Scituate within the compass & energie of the Trades Winds; and they, in
those parts more especially, Blowing almost full West, (as appears by M.ʳ Hally's &

other mapps of the Said Winds;) do, as 'tis naturall, drive the Waters the same way
with themselves; And forasmuch as the Naturall Course of the Tides as appears by
Plate the 3, Figure I. do there Likewise, Set the Same Way with the Said Current: It is
as it were visibly evident, that the force of both these Motions, being thus United
together, must violently drive the Watters against the Land; by which it being Stop'd;
it forthwith Setts to the North; (as you may See it cannot do otherwise,) with that
Rapidity as to cause the Phænomena in the Waters before mentioned: And did not the
Said Winds become calme on the East Side of *Asia*, there would [p. 133] be the like
Phenomena on that Coast, as on the Coast of *America* now described.

The next thing to be noted from Fig: **I.** and **II**,[158] Plate the third; Is the great
quantity of Water, that meets under the Æquinoctial Line; Between 170 and 270,
Degrees of Longitude, (to wit) in the South Sea; compared with the parvity thereof
that meets in the opposite part of the Globe; (Viz:) between 20 and 150 Degrees; it
being there prevented by the **EURO=AFRI=ASIAN** Continent. Now forasmuch as
this inequality in the meeting of the Tydes is Visibly manifest; and Seeing God made
nothing in Vain we will therefore propose what Effect this may have upon the Body of
the Earth: And in the First place it is to be Noted (as in Sect: III.) That Seeing the
Waters are moved at the two Poles of the World, at one and the Same instant and
doubtless, to one and the Same heighth or distance from the Center: That motion,
being Equall, cannot therefore cause the Earth to Ponderate or devolve, either towards
the *north*, or towards the South: But when those Flouds, shall come and meet each other
under the Equinoctiall line, for at least 5000 miles together, in that great ocean, and not
perhaps the Moiety thereof, on the opposite parts of the Earth, (though it riseth near
the Said Line but 2 or 3 foot in heigth;) yet I say, being for So many thousands of miles
together, it must of necessity bring the Earth out of its Æquilibrity; and thereby cause
it to ponderate, either to the East, or West which being constantly repeated, must
unavoidably give it that naturall diurnall Rotation, that has, especially of late years,
been justly ascribed thereunto: For it is evident, that if a Wheell or Sphere, be equally
poized, and its Superficies equally distanced from its Center, as the Earth was in its
aforesaid fluidity; and if after that, any thing be (tho but intermitingly) added to one
Side thereof and not to the other: yet if that addition, be perpetually repeated, it must
of necesity cause it in like manner, constantly to move round upon its center: And the
same may be understood of the Earth, from the causes before [p. 134] mentioned: (And
doubtless if ever the Perpetual motion be found, it must be in imitation thereof.) Only
with this difference, That whatsoever Spherical Body, is suspended in ye open Æther
Unfixed, and therein moved as the Earth is, That primary or centrall Motion, must
unavoidably cause it to Seek a new Center; which being likewise, pepetually repeated.,
must., as Could be easily demonstrated, produce a Secondary, or Orbitall motion;
probably, round the Center or place where it first rested, at the Seperation of the
Chaos before those Orbs were Vivified; as is Shewn in Sect: **I**, in the preceedent
treatice[159]

Having Solved Some of the Phænomena's in the Ocean; and also Shewn that the Disparity, in the Meeting of the Waters in the Æquinoctiall is the Cause of the Earths Rotation; I now come to Explain the Lines P, Æ, P, and P, N, P; Wherein it is to be understood; That forasmuch as the places of the Originall Rising and ultimate Meeting of ye Tides, are 5400. Miles asunder; it cannot reasonably be supos'd, That the Flouds made at either of the Poles, do runn from thence directly to the Said Line, without any Ebbing, in any place, 'till thereunto arrived: I shall therefore, not only Shew, That they do not So pass; but also propose, How many Several Flouds and Ebbs, may, according to the Phænomena's in the Tydes, be at all times existing between the Said places, as they pass on in the open Ocean; And to guide us herein we have these two things that are certainly known.

> First, That the Distance from the respective Poles, to the Æquinoctial, is, as is Said 5400. Miles.

> Secondly, That the certain *Mean* time, for one Ebbing, and once Flowing of the Tydes; is alwayes but 12 Hours and 24 Minuits; (though the difference of Spring & Neap=Tydes is alwayes more or less as by the Tyde Table in Sect: II, appeareth.)

Now forasmuch as one halfe of this time is almost in all places of the World, taken up for the Floud; and [p. 135] the other half for the Ebb; it will evidently follow, that when it has flowed 6. *hours* & 12. *minutes*; it must, as Experience confirmeth, begin to ebb again wheresoever it was then full Sea; though perhaps, in that time it has not run above 6. or 700. miles forward from the place where it was first moved; not in Some places half So much; Whence 'tis plain, from these known truths, that as the flouds do pass from the Poles, to the Line; there must unavoidably, be Severall Flowings and ebbings, following each other, at Some certain Distance, before it can Run the 5400. Miles before Mentioned: Nevertheless tis the *first* Floud, by pushing forward the Ocean, that causeth it to flow on, till 'tis Stop'd either by the Waters meeting each other in the *Line*, or by the land on the respective Coasts; as in the foregoing Treatice is more fully Shewn: And therefore Supposing, as it is but reasonable, that at about every 700. miles Distance, there is an *Ebb* as well as a *Floud*; we may justly represent the Severall Ebbings, & Flowings, as you see them described by the Lines P. Æ. P., & P. N. P, before mentioned; in wch yo^u may conceive the waters in one Latitude, Ebbing towards P, P; And at the Same time in another Latitude Flowing towards Æ, & N: For as Soon as it has flow'd the Said 6. hours & 12. Minutes in any Latitude; it thereupon generally Ebbs, or returns again towards the Poles, or place from whence it came: Or to speak more properly, to the next place where it is become Low water; So that if it be full Sea under the Poles, or in any other Latitude, at noon; In about 6 hours and 12 Minutes after, it will be Low water, in all the Same places; as by comparing the hours, in y^e Said two weaving Lines with any Tide=tables will more fully appear.

Now to prove that these things are not groundless Notions nor Speculative Assertions, but agreeable to Truth & Experience; I have annexed a Compendium of the Flowing of the Tydes, in the English Channel, in Fig: III, Plate 3;[160] by which it appears, That between the *Lands=End* and *Ushant*, it is full Sea at 6 of the Clock (e.g.) in the Morning: But between *Flanders* and the *Downs*, (which is but 300 Miles from the place first mentioned,) 'tis not So, 'till 3 in the Afternoon; the Difference of which time, being 9 Hours; It must (and Experience testifies the Same,) have been Low=water at the first place, and the Tyde returning thither again with halfe=Floud, by Such time as it is full Sea in the *Downs*: Which fully proves, [p. 136] That there must be Several Ebbings and Flowings, before the flouds can pass from the respective poles to the Equinoctiall, because of the great Distances before mentioned; and that the said Compendium is impartially drawn, I refer to the common Tide=tables; as also to Mr *Hally's* accurate map of the English chañel;[161] where you'll find it undeniably verified. Another instance to the same purpose, is what my self observed, when I kept the foregoing Register (Viz^t) That in the *Lake* or *Lagune* Lying between y^e Isle of *Portland*, and the town of *Abbottsbury*;[162] I found it to be full Sea; at the mouth or Inlett of the said Lake, at Least three hours before it was So, at the other end thereof: As by the Said Register appeareth; For the full=sea at the *Inn=lett* is very near the same as at *Weymouth*, th'o the said Lake is not 7 miles in Length in the Whole. Now the reason of this great Difference, in So Short a space; is partly owing, to the winding about of the water, round the said *Island*; but cheifly to the narrowness of the Inlett; that Leasds into the Said Lake: It being not above 4. or 5. perches broad. Forasmuch therefore as it apeareth that the Tides are hastened or retarded in their Course; in Some Proportion according the narrowness of wideness of their Inlets (For we finde that in the first Instance, where the mouth of the Chanel is large, they run about 300 miles in 9 hours which is above ten times faster than in the said Lake.) I therefore reasonably concluded, that in the wide ocean where its Course is not interupted by winding about or otherwise; it might there Flow, 6 or 700 miles in 6 hours and 12 minuts; and ebb as much in the Same time: Which is the reason that induced me to asigne So many miles, for one ebbing and one flowing in the great ocean. From which Instances, together with divers other appearances in Nature, we may reasonably conclude, that in all probability, there is not less than nine or ten Flouds, and as many Ebbs, alwayes Existing between the respective Poles, and Place of meeting; and accordingly we have represented the Same, by ye weaving Lines before mentioned: And tho we cannot pretend fully to demonstrate, their certain number; yet wee can further prove that there are severall [p. 137] of them So Existing, by what we further Experience, in th English Chanel before Mentioned, For you See at the *Entrance* thereof in Fig: III, Plate 3,[163] 'Tis full sea at **XII** a Clock, (e.g.) at Noon; But between *Calis*, and Dover, (by the coming up of that Floud, as the Hours between the said places do manifest,) it is not So, till **XII** at night: Yet at that very **XII** at night, there is another full Sea again, at the said entrance of the channel, which is made by the Floud coming from the *West*; and the Ebb returning from the *East*: And according to this manner of Ebbing and Flowing the Waters are there, and elsewhere, agitated from one full Sea to another; Whence 'tis evident, That there is constantly no less than two fflouds existing, at one and the Same Instant, in the Said Channel, within the Space of 500 Miles; and therefore there must be many more within the distance of 5400 Miles, as was before Asserted.

The next thing intimated by the Said Weaving=Lines, is, That the Tydes or Flouds do loose their Energie; and gradualy diminish their Heighth, and Swiftness in Runing; and also the distance in their approaches to each other, according to the distance, as well in respect of *Time*, as of the *Place*; when and where they are first moved. From whence we may plainly discover the *Cause* why the Waters do generally Flow So faint and inconsiderable, between the Tropicks and near the Æquinoctial; to what they do near the Poles, where they are first moved: Which Faintness M.̲ Dampier,[164] and others, have well observed and entred, in their respective Voyages. For although 'tis Observed, That where the Tydes of floud do pass through any narrow Streight or In=lett, to Supply any large Bay or Lagune; the Flouds at Such In=letts, do comonly Run very Swift, and rise very high; and also flow longer, than what is usual elsewhere: Yet nevertheless at the bottome, or further end of Such Bayes, the Motion is in all these respects quite otherwise: For in *Baffins* and *Hudsons* Bayes, in Lat: 70 deg: North; Where it runs So Swift at [p. 138] the In=lett, as that it riseth 50 Foot parpendicular; and Flows about 9 Hours, and Ebbs three, (both making the mean time before mentioned; as it dos likewise at the Port of *Weymouth*, where 'tis 4 Hours Floud, 4 ho: Ebb, and the remainder dead Water;) Yet I Say, notwithstanding this great Swiftness, and rising at the In=lett, we finde that at the farther end of those Bayes, especially where it Runs to a great distance, as in the Mediterranean Sea, and other places; the Motion becomes So faint and Languid, as that it riseth not above one or two foot in heighth; and in Some places, as in the latter, the motion becomes So Slow, as to be Scarce discernable: The time in Flowing, and Swiftness in Runing, being also diminished accordingly. From whence, as also from the *Logarithmetical* Motion, which I discovered in the Tydes, mentioned in Sect: III, I was taught to make the Weaving Lines P, Æ, P, and P, N, P; [Decreasing] in heighth, and distance from each Floud; as they [increased] in distance, from the time, and place; when, and where, they were first moved: And according to this Weaving forme or Motion, we ought to conceive, all the Waters in the Ocean generally to Flow, or to be alwayes agitated; as before declared.

And it is further to be observed, That if any of these, though but faint Flouds, do flow against, or before the Mouth of any, considerable Rivers It will So Stop their Course as to cause great Risings in Such Rivers though the Waters in the Bays do rise but very little; As appears by the Flowing 15. Foot at Port Nelson on the S:W: of Hudsons Bay, (and divers other Rivers), tho' the Flouds are there So faint as before Observed.

And here it must be also remembred, That none of these Occasionall, or particular
great Risings, or Long flowings; Ought to be regarded, or had in competition, with the
true genuine or naturall Motion of the Tides, in the wide Ocean: For although, (even
towards the Æquinoctial,) there are almost [p. 139] Such Flowings, at Such In=letts; as
at the Lagune of *Trist* in the Bay of Campeche[165] Lat: 17 deg: North; where it riseth 6 or
7 ffoot: And at *S*: *Michael*,[166] on the West Side of America Lat: 5 Deg: North; where it
riseth 20 Foot: As also at the Streights between *Nova Holandia*, and the *Malucca Isles*, Lat
15 Deg: South; where it riseth 30 Foot: Yet I Say, notwith=standing these great Risings,
at the Entrance into Such Bayes, at a Smal distance from them in the wide Ocean,
especially near the Line, it riseth but 2, 3, 4, or 5 Foot; as the Authors before
mentioned, and divers Mariners have observed & Recorded: So that we may be well
assured; That the natural and general Rising of the Tydes, in the wide Ocean, even
near the places where they are *first moved* is not more than about 7 or 8 Foot; and so
down to 3 or 4 Foot as they approach the Æquinoctal Line: as will doubtless more fully
appear if observations be further made, at Some distance from Small Islands in the
midst of the ocean; being the properest places for that purpose.

Nor is the time of full Sea to be truly reckoned, without having respect to the
generall Currant or Setting of the tides as you see them Delineated in the said mapp: For
suppose a S. S. West and N. N. East Moon Should, as is pretended, make Full Sea at
one of the Isles of *Azores*; Latitude 40, Deg: No: and alike S. S. West; and N. N. East
Moon, Should also make Full Sea at the Isle S: *Antonio*,[167] in the Lat: 19 degrees North:
Yet you see, by the course or setting of the Tides, as they are described in the said
Mapps, That it must be Full=sea at the Isles of *Azores*, (If it was full=sea at such moons,)
12 Hours and 24 Minuts; or perhaps 24 ho: 48 before it can be So, [*by that Floud*] at the
Isle S: Antonio: And So more or less, in all other places; as you therein See, the nature
of their Flowing will produce them.

I might here proceed to other *Phanomena's*, as well relating to the trade=winds, as
to the Marine parts of the Earth (for I find none, if *generall*, but what may be accounted
for by the precedent Doctrine:) But I Shall at present only Shew, That the [p. 140]
Waters, by their thus meeting in the *Line*, do fully confute the Old Notion, of their
being Governed by the Moon.

And herein Let us Suppose, That they are Moved by the Moon's pretended
Attraction.

Then it will appear, from the two first Mapps, That whensoever the Moon comes
into the Sign Cancer, (as She dos once in less than every 8. and twenty days,) That
Quantity or moiety of water, which belongs to the South Hemisphere, to wit, from
the South pole to the Equinoctiall; would be drawn 40. or 50. degrees, Further
Northward; by which means, all the Waters between the South Pole and the
Æquinoctiall would not only be greatly attenuated and diminished in its Debth; but

also at the Same time, the other moiety belonging to our North Hemisphere, to wit, from the Æquinoctiall, to the North pole; would, on the contrary, be as much accumulated, and increased in height: From whence it would necessarily follow, that once every Moneth, whether the moon was at full or change, at first or last Quarter; it would make the Tides 2. or 3. times higher, than when, on the contrary, they were drawn to the Tropick of Capricorne; which would be as frequent as the former: Near both which places, it would alternatly remain in that position, for 6. or 8. days together. But inasmuch as this Phænomena, nor any thing like it, is ever found; We may Justly infer, and Assert, that as there is noe Such Effects, there can be no Such Cause; Therefore the moon dos not Attract &c:

Nor is this all the Consequences, that would attend Such a Supposition; For Seeing the dayly Rotation of the Earth as has been Shewn, is caused by the Excess, or inequallity of the waters meeting in the Æquinoctiall, compared with the parvity thereof meeting on the opposite part of the Said Line; It will as naturally follow, That if they Should, in like manner, be made unequall, by being, as is [p. 141] Said, drawn from one Pole towards the other; and made to meet in the Tropicks, as aforesaid, instead of the Æquinoctiall; they would unavoidably bring the Earth out of Æquilibrity in respect of its Poles and thereby, cause, it, alternatly, to turn from North to South, and so from South to North again. The absurdity of both which Suppositions, every Moneths Experience, dos visibly testify.

Wherefore the Waters are not drawn
by the Moon, as aforesaid.

And what is here Said of its attractions, may be also understood, Mutatis Mutandis, of its Supposed Pressure.

From both which, observations, being So fully confirmed by Experience, 'tis Sufficiently evident, That tho there is, as has been Said, Some Conformity in respect of [*Time*] between the Motion of the Moon, and that of yᵉ Tides; Yet I Say notwithstanding this; The Moon has no power or Influence to draw or Attract; nor by pressure, to Repel or drive back, the least drop or part of yᵉ Ocean, in any manner whatsoever. For although a thousand Clocks, and as many Watches, may in like manner agree with each other, as is Said, in respect of *time*; Yet we all know, That neither of them is the *Cause* of the others Motion.

For a Conclusion therefore, Seeing it appears by the I, and II, Fig: Plate 3;[168] (Or at least by the Volumns from whence they were derived,) That the Tydes do Originaly proceed from the two Poles of the World: And that the mean Time for one Floud, and one Ebb, jointly taken, is but 12 ho: and 24 Minuts: And Seeing [p. 142] also the Tydes do certainly require a constant Resusitation or lifting up, to continue their Motion of Ebbing and Flowing: What may we Infer from these and many other appearances in Nature: (for the very invisible things of God, are manifest, by the things that are made and do appear,) but that the Waters are at first, by a kinde of Systolation, and

Diastolation, gradualy thrust up or put in Motion *at* and *round about* the two Poles of the World, as we have before Asserted.

Secondly, That the Semi=Diameter of the Quantity of Water that is there first moved, before it begins to Subside again, is Somewhat more than the aforesaid Distance of its flowing in the Ocean, when it comes under our Latitude (Viz:ᵗ) above 700 Miles; because it runs faster & farther at the places and times, of its frist Moving than it does afterwards: Nevertheless the Semidiameter of the Musculous parts of the Earth that gives Motion thereunto, are not respectively, perhaps one fifth or one tenth of that Measure.

Thirdly That the time Spent in its frist Raising up, is not only less than the mean time of 6 ho: and 12.^{Min} But also less than – – – 6 ho: and 7.^{Min} Because you may See in the frist table beforegoing; the Distance in flowing at Spring=tides; is but 28^{Min} in 24. hours which is for two flouds and two Ebbings; and therefore it can be but one fourth part thereof (Vizᵗ) 6 ho: and 7.^{Min} for the bare flowing or raiseing up of one floud – – – when it comes on our Coast: and Consequently, it must be less than that [p. 143] where it is frist moved; as the nature of their Flowing dos Demonstrate.

Fourth, that the *like time* is Spent in its Sinking down again.

Fifthly that this motion is constantly repeated or renewed, at the End of every 29 Ebbings and as many Flowings there being no more in the 15 Days that are comonly reckoned from one Spring=tide to another.

Sixthly that the time when this Lifting up is first begun, is about 31. Hours before they begin to lift; (as 'tis rightly called) here; (which the Hours in the first Weaving Line at our Latitude do plainly intimate;) Namely, When the *Moon* is about 8. Dayes & $\frac{1}{2}$. and 23. Dayes and $\frac{1}{2}$ Old.

Lastly, that by and from these two first Flouds, and the Energie thereof; All the Ebbings and Flowings in yᵉ Oceans, are Constantly agitated and produced, and thereunto Justly conformed Assimulated and continued.

Finis.

Notes and references

My practice in these notes has been first and foremost to identify proper names, places, and references to works cited by Hobbs. To some degree the notes contain a commentary on the more geological aspects of Hobbs's thought. I have thought this worthwhile because discussion of the crust of the Earth is the central subject of Hobbs's tract, and is the most intellectually interesting feature of his work. I have not attempted an analytical commentary on Hobbs's broader astronomical and cosmic speculations.

The recto of the first blank sheet of the MS before the title page carries the following note in ink

> J. Bailey
> August 1825
> I bought this MS in the borough of Southwark. Shortly after I had made the purchase mention was made of it, The Title quoted partially &c, in the Times newspaper.
> J.B.

Also on this page in ink there is the jotting
> Parkinson's Organic Remains of a former world, 3 vols.
> Sowerby's Conchology 4 vols

and in pencil

> The Earth Generated
> *Britscha* nearly new, complete & for ready use. Is only to be seen at 93 Long Acre opposite Bow St Times Sept 19 1825

I have not been able to ascertain with certainty the identity of John Bailey. James Parkinson's *Organic remains* was published in three volumes between 1804 and 1811 (London). James Sowerby's *The mineral conchology of Great Britain* appeared in seven volumes between 1812 and 1846. A Britscha is an open-air carriage.

TITLE PAGE
The five words which constitute the main title are on a slip of paper pasted over the original title, which is
'The Generating & Anatomizing the Earth'

1 The complete verse—in the King James version—of Isaiah 5: 12 runs: 'And the harp, and the viol, the tabret, and pipe, and wine, are in their feasts: but they regard not the work of the Lord, neither consider the operation of his hands'.

The complete verse of Romans 1: 20 is: 'For the invisible things of him from the creation of the world are clearly seen, being understood by the things that are made, even his eternal power and Godhead; so that they are without excuse'.

The complete verse of Wisdom 7: 17 is: 'For he hath given me certain knowledge of the things that are, namely, to know how the world was made, and the operation of the elements'.

Hobbs seems to have used the King James Bible, but does not always quote completely accurately.

2   It is worth noting that Hobbs nowhere gives a technical name to the science he is pursuing.

3   It is unclear which counties. The only ones mentioned by name are Dorset, Devon and Cornwall. Hobbs's experience of strata seems to have been chiefly of sedimentary rocks, from which one can assume chiefly a familiarity with the geology of the more southerly and easterly parts of the British Isles.

4   Figures 1, 3, 4, 8, are all missing. They are here conjecturally reconstructed at the end of this edition.

5   Hobbs's remark is perceptive, though it is strange that he says he has met it nowhere, since it is a dictum of John Woodward, with whose *Essay towards a natural history of the earth* (London, 1695) Hobbs was obviously familiar.

6   Note that Hobbs has no technical, generic term for 'fossils'. He never uses terms such as 'extraneous fossil' or 'formed stones'.

7   Hobbs is here most likely referring to the coastline between Weymouth and Poole. *Cf.* G. M. Davies, *The Dorset coast; a geological guide* (London, 1935); A. W. Rowe, *The white chalk of the Dorset coast* (London, 1901); W. J. Arkell, *The geology of the country around Weymouth, Swanage, Corfe and Lulworth* (Geological Survey Memoir) (London, 1947).

8   Figure 2 is lost, and has been reconstructed.

9   This is typical of the homely, popular metaphors which Hobbs uses throughout. It also embodies his overall alchemical interest in processes such as fermentation and baking.

10   Hobbs here summarizes Genesis 1.

11   Note the omission of any mention of Man. The entire tract is remarkable in respect of how little it discusses the history and functions of the Earth in terms of Man.

12   Genesis 2: 4 reads: 'These are the generations of the heavens and of the earth, when they were created, in the day that the Lord God made the earth and the heavens'.

Hobbs's philosophy of generation fused together two main currents of seventeenth century scientific thought. On the one hand there was a tradition of analysis of living bodies, which would include William Harvey's *Exercitationes de generatione animalium* (London, 1651); Sir Kenelm Digby's *Two treatises, of bodies* (London, 1650, see especially pp. 201–65, 290–302); N. Highmore, *The history of generation* (London, 1651). For commentary on these see Joseph Needham, *A history of embryology* (second edition, Cambridge, 1959), pp. 120–39; Philip C. Ritterbush, *Overtures to biology* (New Haven and London, 1964); Elizabeth B. Gasking, *Investigations into generation 1651–1828* (London, 1967, especially pp. 56–65); T. S. Hall, *Ideas of life and matter*, vol. 1 (Chicago and London, 1970).

At the same time, there was also a tradition of comprehensive alchemical philosophy, which interpreted Nature in terms of processes such as generation, organization, distillation, fermentation, corruption and putrefaction. For some guidance to seventeenth century works in this tradition see B. J. T. Dobbs, *The foundations of Newton's alchemy* (Cambridge, 1975, especially pp. 44–60 and bibliography); Allen G. Debus, *The English Paracelsians* (London, 1965); Owen Hannaway, *The Chemists and the Word* (Baltimore and London, 1975); Walter Pagel, *The religious and philosophical aspects of van Helmont's science and medicine* (Supplement to the *Bulletin of the history of medicine*, no. 2, Baltimore, 1944).

Here, as elsewhere, it is very difficult to grasp how widely Hobbs had read in these traditions. He occasionally quotes directly (*e.g.*, from Digby and Basil Valentine), but his overall philosophy is an eclectic amalgam, not directly or totally dependent on any one source. As a former excise officer, Hobbs must have had considerable first-hand experience of brewing, distilling and refining. This might help to account for his chemical philosophy of the world.

13 This kind of interest in God's creativity is characteristic of seventeenth century exponents of alchemical philosophy such as Robert Fludd and Gabriel Plattes. *Cf.* the works by Debus and Hannaway cited in the previous note, and Debus's 'Gabriel Plattes and his chemical theory of the formation of the earth's crust', *Ambix* **9** 1961: 162–5.

14 In other words, Hobbs was claiming that, as in the formation of a foetus all parts generate and grow simultaneously, so likewise the formation of different features of the Earth is a synchronous process. The 'days' of Creation are not successive stages in time, but rather different aspects of an organic, synchronic whole, 'a common Coagulation, and General Graduation'.

This is a radical solution to the issue of the order of the Mosaic days. Not only does it forestall the problems which arise from treating the days literally as periods of twenty-four hours. It also avoids the difficulties which ensue from the Biblical order of creation, in which for instance land and sea are supposed separated before the creation of life, thus making it obscure for many naturalists to explain how fossils had become embodied in the rocks.

15 Note Hobbs's stress on the fecundity of Nature. Here as elsewhere Hobbs stressed that it was Nature which brought forth creatures, not God directly and miraculously. Hobbs strenuously defended spontaneous generation. For late seventeenth century debates on the validity of spontaneous generation see John Farley, 'The spontaneous generation controversy (1700–1860): the origin of parasitic worms', *Journal of the history of biology* **5** 1972: 95–125; E. Mendelsohn, 'Philosophical biology *vs* experimental biology: spontaneous generation in the seventeenth century', *Actes du XIIe Congrès Internationale d'Histoire des Sciences* **1** 1968: 201–26; Edward J. Foote, 'Harvey: spontaneous generation and the egg', *Annals of science* **25** 1969: 139–63; F. J. Cole, *Early theories of sexual generation* (Oxford, 1930). For a contemporary discussion see Matthew Hale, *The primitive origination of mankind* (London, 1677, section iii, ch. iv, 266f.).

Hobbs's belief that it was part of the essential nature of matter to be productive of life was contrary to the trend of the mechanical philosophy of his day—which he clearly opposed. *Cf.* M. C. Jacob, *The Newtonians and the English revolution 1689–1720* (Hassocks, Sussex, 1976); R. L. Colie, *Light and Enlightenment* (Cambridge, 1957).

16 Psalm 33: 6. 'By the word of the Lord, were the heavens made; and all the host of them by the breath of his mouth'.

17 The passage from 'As to the fact' down to 'the Judicious Reader' has been deleted in the MS. The two undeciphered words do *not* appear to be 'Thomas Hobbes'. Presumably Hobbs means 'neighbour' both in the sense of geographical vicinity—Thomas Hobbes hailed from Malmesbury in Wiltshire—and also in the sense of the similarity of names. William Hobbs's disclaimer of any relation with Thomas Hobbes shows the continuing reputation of the latter as a bogey-man. The deleted passage reflects Hobbs's deep awareness that he was an unknown provincial scholar.

18 In his letter to the Royal Society (read 11th May 1709), Hobbs admitted his deficiency in learning in languages. In the present work he cites a few Latin tags. There is no conclusive evidence whether he could read Latin fluently and extensively.

19 From 'know in musick' to 'never played' is deleted.

20 The 47th problem is Pythagoras's theorem.

21 Hobbs was probably here thinking of Bishop Virgil of Salzburg (710–84), whose unorthodox views about the Earth (*'quod alius mundus et alii homines sub terra sunt'*) have often been construed in this sense. It is not clear exactly what the bishop had in mind. For his views he was upbraided, but neither deposed nor put to death. *Cf.* G. Sarton, *Introduction to the history of science*, vol. i (Baltimore, 1927: 516).

22 For Archimedes's work on water displacement and specific gravity see E. J. Dijksterhuis, *Archimedes* (Copenhagen, 1956: 18). The story of Archimedes running through the streets without his clothes was passed down in Vitruvius's *De architectura*, book ix.

23 John Napier, 1550–1617, inventor of logarithms.

24 Hobbs seems here to overestimate his own originality in matters relating to the strata, in which the work of Woodward—with which he was familiar—and Steno had certainly paved the way. This remark may perhaps indicate Hobbs's relative lack of familiarity with scientific treatises on the Earth already published.

25 William Lilly, 1602–81, was the most famous and prolific astrologer of his day, for whom see D. Parker, *Familiar to all: William Lilly and astrology in the seventeenth century* (London, 1975). It is not clear precisely to which of his treatises Hobbs was here referring. Astrology still had many supporters amongst naturalists in England in the latter part of the seventeenth century, such as Hobbs's fellow Dorset man, Joshua Childrey (see the Preface to his *Britannia Baconica* (London, 1661)) and John Aubrey (*cf.* Michael Hunter, *John Aubrey and the realm of learning* (London, 1975: 117–30, 140–6). For attacks on astrology see K. Thomas, *Religion and the decline of magic* (London, 1971) and John Redwood, *Reason, ridicule and religion* (London, 1976: ch. 6).

26 Hobbs's attack on astrology is consonant with his opposition to all forms of action at a distance, as being occult and lacking foundation in Nature.

27 'There are as many opinions as there are men.'

28 The words 'Weymouth' and 'William Hobbs', and the '15' of 1715 have been scratched out.

29 This head calls us back to the original title of the work.

30 Hobbs's opinion that Scripture was not to teach natural philosophy was a more liberal view than was common amongst such contemporaries as John Woodward and John Keill (1671–1721).

31 In his letter to the Royal Society (read 11th May 1709) Hobbs specifically stated 'above thirty years'.

32 Hobbs cited rather few authors by name. He seems to have had a general knowledge of currents and natural philosophy in the second half of the seventeenth century, lacking, perhaps, a sharp and critical intellect, particularly in his discussions of the more mathematical sciences.

33 'Fossils' here means rocks or minerals in general.

34 In suggesting that coal contains no organic remains Hobbs seems to show he was ignorant of the work of John Beaumont, whose papers had pointed to the presence of plant fossils in coal. *Cf.* 'On rock plants, and their growth', *Phil. trans R. Soc. Lond.* **9** 1676: 724–42; *idem*, 'A further account of some rock plants of Mendip Hills', *Phil. trans R. Soc. Lond.* **13** 1683: 276–9. Hobbs's deep interest in the transformation of sea-bed into solid land was obviously strengthened by his actual observations of processes occurring on the Dorset coasts.

35 Hobbs was presumably highly familiar with the Portland quarries, for which see F. H. Edmunds and R. J. Schaffer, 'Portland stone, its geology and properties as a building stone', *Proceedings of the Geologists Association* **43** 1932: 225–40: A. M. Wallis, 'The Portland Stone Quarries', *Proceedings of the Dorset Natural History and Archaeological Society* **12** 1891: 187–94; W. J. Arkell, *The geology of the country around Weymouth, Swanage, Corfe and Lulworth* (London, 1947); J. C. Mansell-Pleydell, 'Geological notes on the Isle of Portland', *Proceedings of the Dorset Natural History and Archaeological Society* **6** 1884: 58–65; W. J. Arkell, 'The Names of the strata in the Purbeck and Portland stone quarries', *Proceedings of the Dorset Natural History and Archaeological Society* **66** 1944: 158–68. Sir Christopher Wren was highly familiar with the Portland stone quarries—much of the rebuilding of London after the Great Fire of 1666 was accomplished using Portland stone, which Wren championed. Wren became M.P. for Melcombe Regis in 1701. It is interesting to speculate whether Hobbs may have met Wren. See B. Little, *Sir Christopher Wren, A historical biography* (London, 1975: especially pp. 199–232).

36 Proverbs 8: 24, 25: 'When there were no depths, I was brought forth; when there were no fountains abounding with water. Before the mountains were settled, before the hills was I brought forth'.

37 Figure 1 has been lost, and is here reconstructed.

38 Figure 3 is lost, and is here reconstructed. Hobbs's observation that there is a conformity between relief and the angle of the strata is one presumably stimulated by the scenery of east Dorset, and perhaps especially its cliffs.

39 Figure 4 is lost, and is here reconstructed.

40 This passage is a little obscure, perhaps largely because of the loss of the original diagrams. Hobbs's main point seems to be that wherever there are hills, there has been *elevation*; and elevation entails that at certain points—*e.g.*, the exposed scarp edges of hills—the strata have been visibly fractured. He was denying the possibility of what is pictured in Figure 5, *i.e.*, the line of strata everywhere perfectly following the contours of the countryside.

41 Figure 5 is lost, and is here reconstructed.

42 Figure 2 is lost, and is here reconstructed.

43 In 1709 Hobbs sent some 'models' to the Royal Society. These included a map of the Isle of Portland. It is now lost. For the properties of Portland stone see J. Smeaton, *A narrative of the building of and a description of the construction of the Eddystone lighthouse with stone* (London, 1791).

44 Hobbs recognized that no ordinary deluge of water could conceivably raise marine shells and deposit them deeply within already existing consolidated strata. Nor could the Biblical Deluge have had that effect, since it could have no power to soften or dissolve the rocks and admit the fossils. Hobbs's statement is almost certainly an attack on John Woodward's *An essay towards a natural history of the earth* (London, 1695).

45 This is not a commonplace contemporary perception. For example there is no mention of this device in J. C., *The compleat collier* (London, 1708). Hobbs's point is however made in considerable detail in George Sinclair, *The hydrostaticks* (Edinburgh, 1672: esp. 258–80).

46 The division of matter into these three principles is an essential part of the Paracelsian tradition of chemistry and alchemy. *Cf.* A. G. Debus, *The English Paracelsians* (London, 1965). For a roughly comparable attempt to integrate an alchemical philosphy of organic creation harmoniously with the Bible cosmogony see the work of Thomas Robinson (d. 1719): *The anatomy of the earth* (London, 1694), and *An essay towards a natural history of Westmorland and Cumberland* (London, 1709).

47 The Bible quotation is from Genesis 1: 2: 'And the Spirit of God moved upon the face of the waters'; v. 7: 'And God made the firmament, and divided the waters which were under the firmament from the waters which were above the firmament; and it was so'. For alchemical interpretations of Genesis, *cf.* note 46.

48 For the works of the so-called Basil Valentine see the article (*sub* VALENTINE) by A. G. Debus in the *Dictionary of scientific biography* **13** 1976: 558–60. Basil Valentine's work is a compendium of sixteenth century and seventeenth century alchemical ideas. Hobbs is quoting from *The last will and testament of Basil Valentine* (London, 1670). The quotation is accurate, except for minor errors of punctuation, etc. Part of this work is readily accessible in *The Hermetic Museum restored and enlarged*, ed. A. E. Waite, 2 vols (London, 1893 **1**: 330–45).

49 Job 10: 10: 'Hast thou not poured me out as milk, and curdled me like cheese'.
The epitome is of course Man.

50 Proverbs 8: 24, 25; here quoted for a second time.

51 The quotation is from Sir Kenelm Digby, *Two treatises, in the one of which the nature of Bodies; in the other, the nature of mans soule; is looked into, in way of discovery of the immortality of reasonable souls* (London, 1645: 275). The quotation is substantially accurate. Many of Hobbs's ideas bear a quite close resemblance to Digby's beliefs about the chemistry of bodies, as set out in this work. For an account of Digby's life and work see R. T. Petersson, *Sir Kenelm Digby, the ornament of England, 1603–1665* (London, 1956). For interpretation of Digby's experiments, see Needham, *op. cit.* (note 12) pp. 124–7, 135–40.

52 *I.e., The works of the Hon. Robert Boyle, Esq., Epitomiz'd by Richard Boulton* (3 vols, London, 1699–1700, **I**: 73). Hobbs's quotation is substantially accurate.

53 This is certainly a jellyfish, most probably *Aurelia aurita.* The whiter parts are probably the sex organs. The dilations and contractions are its mode of propulsion. I have not come across any other usage of the local term 'Bulls'. Hobbs sent a drawing of 'Bulls' to the Royal Society in 1709. It has been lost.

54 Figure 6 is lost, and is here reconstructed.

55 The Euclid reference can be found in *Euclid's Elements of geometry* published by John Leeke and George Serle (London, 1661), 110 thus: 'If in a circle ABC, two right lines AB and CD divide one another (in the point E), the Rectangle contained under the two parts of the one AE and EB is equal to the Rectangle conteined under the two parts CE and ED of the other'.

56 For histories of spontaneous generation, see above, note 12.

57 This represents a clear attack on the beliefs of John Woodward as set out in his *Essay towards a natural history of the earth* (London, 1695). Woodward's belief that strata were arranged in order of their specific gravity met almost uniform opposition. It is not clear what Hobbs has in mind a little lower down when he asserts that beds of chalk etc. are found under mines or beds of lead, tin, silver, etc.

58 This is an important juncture of Hobbs's philosophy of the Earth. Most writers who accepted the organic analogy of micro- and macrocosm attributed to the Earth, as to other living bodies, hair, veins, bones, warts, a mouth, an anus, etc.—as in particular did Hobbs's contemporary Thomas Robinson in his *The anatomy of the earth* (London, 1694). Hobbs perceptively rejected this one-to-one equivalency of physical and physiological attributes. For Hobbs, the organic analogy served chiefly to guarantee the belief that the Earth's economy was an integrated system, with each part functional to the sustenance of the whole. Generally on such analogies see G. L. Davies, *The earth in decay* (London 1969: ch. i). For Robinson, see F. J. North, 'The Anatomy of the earth—a seventeenth century cosmology', *Geological magazine* **71** 1934: 541–7.

59 Job, 26: 7: 'the Lord hath stretched out the North over the empty places, and hanged the Earth upon Nothing'.
 Hobbs's conviction that the Earth was a perfect, self-contained economy—a belief he carried even to the length of asserting that the Earth's intrinsic heat was far more important than the Sun's heat in the generation and sustenance of terrestrial life—was unusual for this period. Far more common was the view either that the Earth was senescent and in decay, or that it had been in some way flawed by the Universal Deluge. Most contemporaries of Hobbs believed that the Earth would in time fall into an uninhabitable condition. Hobbs on the other hand stressed the perfect organic adaptation of the Earth, a view which was hardly found till James Hutton's vision of the Earth as a scene of 'life, death and circulation'.

60 *I.e.,* the great whale or sea monster referred to in Job 41: 1; Psalm 74: 14; Psalm 104: 26; Isaiah, 27: 1.

61 One of Hobbs's rare references to the place of man on the scale of existence and in the scheme of history. Hobbs was almost silent on human destiny, or on the relation of Man to the rest of Creation as set out in the Bible.

62 Job 42: 7: 'And it was so, that after the Lord had spoken these words unto Job, the Lord said to Eliphaz, the Temanite, My wrath is kindled against thee, and against they two friends, for ye have not spoken to me the thing that is right, as my servant Job hath'.

63 For theories as to how the Earth generates minerals, see F. D. Adams *The birth and development of the geological sciences* (reprint, New York, 1954: chs iv, v, vi, ix). Hobbs's ideas here somewhat resemble those of Gabriel Plattes's *A discovery of subterraneall treasure* (London, 1639) (for which see A. G. Debus, 'Gabriel Plattes and his chemical theory of the formation of the earth's crust', *Ambix* **9** 1961: 162–5), and Thomas Robinson, *The anatomy of the earth* (London, 1694).

64 For a guide to contemporary lunar theory, see I. B. Cohen, *Isaac Newton's "Theory of the Moon's Motion" with a biographical historical introduction* (Dawson's, Folkestone, 1975).

65 Figure 10 survives.

66 'Magnetic' theories of the pull of the Moon on the tides were quite common up to the late part of the seventeenth century, being associated with such writers as William Gilbert, Simon Steven, Johannes Kepler and Joshua Childrey. For a contemporary survey of theories of the tides (and an important contribution in its own right) see John Wallis, 'An essay exhibiting the hypothesis about the flux and reflux of the sea', *Phil. trans R. Soc. Lond.* **1** 1666: 263–89. For modern assessments of the range of tidal theories available see M. Deacon, *Scientists and the sea* (London, 1971: 93–116), and E. J. Aiton, 'Galileo's theory of the tides', *Annals of science* **10** 1954: 44–57; and also the article by John D. North on Childrey in the *Dictionary of scientific biography* **3** 1971: 248. Childrey lived at Upwey, half a dozen miles from Weymouth; but since he died in 1670 it is unlikely that he knew Hobbs personally, unless Hobbs was very old when he wrote his treatise. For Childrey's comments on the unusual tides of the Weymouth area see A. R. and M. B. Hall, eds, *The correspondence of Henry Oldenburg*, **5** 1668–9 (Madison, 1968: 384–6 and (on tide theory) 455–6).

  The Royal Society's *Philosophical transactions* carried an important debate on tidal theory during Hobbs's lifetime, as well as publishing tide tables. Newton's theory of the tides was made most easily available by Halley in 'The true theory of the tides', *Phil. trans R. Soc. Lond.* **19** 1697: 445–57. Whether or not Hobbs read such articles actually in the *Philosophical transactions*, we do know—from his citation in his 'Essay concerning Motion'—that he had read the 1705 edition of Halley's *Miscellanea curiosa*, in which many of these papers were reprinted.

67 It is unclear here whether Hobbs was here specifically attacking Newton, or campaigning generally against all forms of action at a distance envisaged as quasi-astrological.

68 For Halley's views on the Atmosphere, see the following articles in the *Phil. trans R. Soc. Lond.*: 'An historical account of the trade-winds and monsoons', **16** 1686: 153–68; *idem*, 'An account of the circulation of the watery vapours of the sea', **16** 1687: 468–73: 'An account of Dr. R. Hook's invention of the marine barometer, with its description and uses', **22** 1701: 791–4; and see A. Armitage, *Edmond Halley* (London, 1966: 75–83). See also the informative article, 'Atmosphere', in John Harris's *Lexicon technicum*, 2 vols (London, 1708–10). Halley's computations of the height of the 'sublimations' of the Earth (45 miles) are to be found in his 'A discourse of the rule of the decrease of the hight of the mercury in the barometer', *Phil. trans R. Soc. Lond.* **16** 1686: 104–16. For a further discussion of Hobbs's use of Halley's ideas on the atmosphere, see below note 92.

69   For Huygens see A. E. Bell, *Christiaan Huygens and the development of science in the seventeenth century* (London, 1947: 161–4), and the article by H. J. M. Bos in the *Dictionary of scientific biography*, **6** 1972: 610. It is almost certain that Hobbs obtained his information about Huygens by reading the English translation of his *Cosmotheoros*, i.e., *The celestial worlds discover'd* (London, 1698: *cf.* p. 131 ff).

70   Further evidence that Hobbs saw all forms of action at a distance as instances of phoney astrological powers.

71   *I.e.*, Saturn, Jupiter and Mars.

72   Almanacs carrying tables of tides appeared annually in Hobbs's lifetime. Flamsteed published a good many in the *Phil. trans*.

73   Hobbs's register of the tides is one of the most impressive surviving provincial records of the tides of its time, for by the early eighteenth century the enthusiastic wave of tide recording which had followed on the founding of the Royal Society had abated somewhat (see M. Deacon, *Scientists and the sea* (London, 1971: ch. v). It fully justifies Derham's comment to the Royal Society: 'Particularly I think his Observations of the Tydes may well Deserve the Cognisance of the Society, he having Observed them strictly for two years, and no doubt will if desired proceed with Delight in further Observing them if it be thought necessary'. The form in which Hobbs kept his register is standard enough. The gaps in the record might be explained by prolonged visits which Hobbs had to make on his Excise business. The purpose for which Hobbs kept the register was to demonstrate empirically that the retardation of the time of high tide was not by a fixed number of minutes per day after full moon, but rather by a progressively increasing interval. He sought thereby to show that the intervals of the tides were not governed by the moon, but were governed by wave motions set up by the pulse of the 'heart' of the Earth every fifteen days, motions that gradually diminished in speed and intensity. In fact, tide patterns in the part of the English Channel around Weymouth are so complex as to render any general deductions from Hobbs's record meaningless.

74   *I.e.*, the River Wey. For a brief account of the geography of the area see R. Good, *Weyland: the story of Weymouth and its countryside* (Dorchester, 1945).

75   In other words, Hobbs reversed the normal causal arrow, and suggested that the rotation of the Moon was governed by the motion of the tides, not *vice versa*.

76   *I.e.*, Thomas Burnet, *Telluris theoria sacra* (London, 1681 and 1689; English translation, 1684 and 1690).

77   Hobbs was here developing a view of the relation of the Moon to the Earth which depends ultimately on a Cartesian vortex theory. The theory as set out by Hobbs, however, certainly is not Descartes's. See E. J. Aiton, *The vortex theory of planetary motions* (London, 1972).

78   *I.e.*, Kepler's second so-called law.

79 For contemporary theories of winds see E. Halley, 'An historical account of the trade-winds and monsoons', *Phil. trans R. Soc. Lond.* **16** 1686: 153–68; E. G. R. Taylor, *The haven-finding art* (London, 1971: 239–40); A. Armitage, *Edmond Halley* (London, 1966: 83–6), and C. A. Ronan, *Edmond Halley, genius in eclipse* (London, 1969: 79–80). Hobbs's notion that winds are a product of the combination of the Earth's rotation with terrestrial obstacles is quite similar to Galileo's theory, and distinct from those of Hooke and Halley, who attributed winds much more to differentials in density and temperature.

This passage is a good example of the transformation of Hobbs's thought between the time of the papers submitted to the Royal Society and this present treatise. In the former, his thoughts on trade winds, etc., were presented as a relatively self-contained speculation. Here they have been organized and subordinated within a more comprehensive and digested view of the relationship of the surface of the Earth to the Moon and the tides.

80 Hobbs's views on madness once again convey his hostility to all kinds of astrology. For the following popular superstitions regarding the Moon see J. Brand, *Observations on popular antiquities, with the additions of Sir Henry Ellis* (new imp., London, 1913: 657–63); Keith Thomas, *Religion and the decline of magic* (London, 1971: 239, 296–7, 333–5, 610, 620, 649); R. Hunter and I. Macalpine, *Three hundred years of psychiatry, 1535–1860* (London, 1963: 285, 459, 480); and C. Plinius Secundus, *The historie of the world*, translated by Philemon Holland (London, 1635, **1**: 163, 488–9; 44; **2**: 397). Such beliefs had recently been satirized in Samuel Butler's *Hudibras*.

For a contemporary survey of theories of the cause of menstruation in women see J. Freind, *Emmenologia* (London, 1729; first, Latin, ed., Oxford, 1703: ch. ii, pp. 4–13).

81 For magnetic theories of the tides, see note 66.

82 A reference to the Cartesian theory of the tides, for which see M. Deacon, *Scientists and the sea* (London, 1971: 50, 52).

83 Figure 11 is lost, and has been reconstructed.

84 For Torricelli's (1608–47) experiment see the article by Mario Gliozzi in the *Dictionary of scientific biography* **12** 1976: 438.

85 Hooke and Boyle were both involved in experiments of this kind. *Cf. The works of the hon. Robert Boyle* epitomiz'd by Robert Boulton (London, 1709, **1**: 384–5), and E. Mendelsohn, *Heat and life* (London 1964: 52–3). It is very likely that Hobbs's source here was J. Drake, 'A discourse concerning some influence of respiration on the motion of the heart hitherto unobserved', which he would have read in Halley's *Miscellanea curiosa* (3 vols, London, 1705, **1**: 173–204, 187–90.)

86 For a contemporary example of a diving bell, see the work of Halley, especially 'The art of living under water: or a discourse concerning the means of furnishing air at the bottom of the sea, in any ordinary depths', *Phil. trans R. Soc. Lond.* **29** 1716: 492–9. Hobbs is of course mistaken in believing that contemporary diving-bells cut off the pressure of the atmospheric column.

87 I have not been able to trace Hobbs's 'Monsiur'. For the development of the barometer, see W. E. Knowles Middleton, *The history of the barometer* (Baltimore, 1964).

88 Measuring chains had been standardized in seventeenth century England. The most famous was Gunter's chain which was made of hard wire, and was four poles in length, consisting of a hundred links. It was used for surveying. The provision of a count-wheel device for a coach to measure distance travelled has a long history, being traceable back to Vitruvius. See A. W. Richardson, *English land measuring to 1800; instruments and practices* (Cambridge, Mass., and London, 1966: 109) and, for a contemporary account, James Moxon and Thomas Tuttell, *A mathematical dictionary* (London, 1700), which is an appendix to Joseph Moxon, *Mathematics made easie* (3rd ed., London, 1700).

89 *I.e.*, Figure 10.

90 Dr N. may be Dr William Nicholls, D.D. (1664–1712), a prolific theological writer in the deist disputes of the late seventeenth century, who discusses most of the problems of contemporary theories of the earth in his *A conference with a Theist, wherein are shewn the absurdities in the pretended Eternity of the World 2. The Difficulties in the Mosaick Creation are cleared 3. The Lapse of mankind is defended, against the Objections of Archaeologiae Philosophicae, The oracles of Reason, etc.* (London, 1696). This work however, does not contain references to rainbows.

It is much more likely, however, that Hobbs intends Sir Isaac Newton, though Hobbs had called Newton 'Sir Isaac' in his earlier letter to the Royal Society. Newton was never 'Dr' (Hobbs's usage *may* be ironic), and it is not clear what Hobbs means in calling Newton a 'reconciling Naturillist'. Hobbs's close association of gravity and the rainbow strongly suggests Newton (whose *Opticks* had appeared in 1704). Hobbs cannot have been unaware that his ideas differed markedly from those of Newton, and he may have felt some animus against Newton on account of his slighting treatment by the Royal Society. Hobbs may well have read 'A letter of Mr Isaac Newton, Professor of the Mathematicks in the University of Cambridge, containing his new theory about light and colours' in Halley's *Miscellanea curiosa* (London, 1705, I: 97–117), where rainbows are mentioned p. 109.

91 The idea that the changing distance of the Moon from the Earth (as between the apogee and perigee) significantly altered the power of the tides is largely connected with Joshua Childrey. *Cf.* M. Deacon *Scientists and the sea* (London, 1971: 103–4) and Childrey's *Syzgiasticon instauratum* (London, 1653) and his *Britannia Baconica* (London, 1661: 96–8). Newton elaborated the idea (Deacon note 109). Notwithstanding Hobbs's ignorance and misapprehensions about the philosophy of the tides, he is of course justified in pointing out that the harmonies of the tides in any particular location in fact bear little close and immediate relation to the phases of the moon. For contemporary lunar theory see E. G. Forbes, *Greenwich Observatory*, 3 vols (London, 1975), vol. i, *Origins and early history*, 1675–1835; and *idem*, ed., *The Gresham Lectures of John Flamsteed* (London, 1975).

92 Figure 12 survives. For an experiment which may have suggested Hobbs's feather experiment see Sir Kenelm Digby, *Two treatises* (London, 1645 ch. viii, 74–6). The suppositions in the following paragraphs concerning atmospheric pressure and the density of the aether are somewhat arbitrary and difficult to follow. Many of the data on which they depend are to be found in E. Halley, 'A discourse of the rule of the decrease of the hight of the mercury in the barometer', *Phil. trans R. Soc. Lond.* **16** 1686: 104–16. Hobbs seeks to use Halley's work to support his claim that air pressure would be utterly feeble at the edge of the atmosphere. His argument is that water is 840 times denser than the air at the surface of the earth. A column of *this* air 45 miles high would contain 840 × 6700 gallons; so the air at 45 miles high is 6700 times less dense than water. But water is incompressible, whereas air is very compressible, and aether must be even less dense than air. Hence the air at 45 miles is very feeble indeed.

93   Figure 11 survives.

94   See note 91.

95   Figure 9 has survived.

96   The seventeenth century saw, of course, a massive debate on the place of the heart in the organism. *Cf.* for example W. Pagel, *William Harvey's biological ideas* (Basel, 1967); G. Whitteridge, *William Harvey and the circulation of the blood* (London, 1971); E. Mendelsohn, *Heat and life* (London, 1964), and for a fundamental discussion of seventeenth century theories of life and generation which serves as a background to this entire section, H. B. Adelmann, *Marcello Malpighi and the evolution of embryology* (5 vols, Ithaca, 1966: especially vol. ii, pp. 752–81).

97   The orbis minor is a sea urchin probably *Echinus esculentis*. Note how the very name echoes the macrocosm/microcosm analogies which are central to Hobbs's thought. For a contemporary account see N. Grew, *Musaeum Regalis Societatis* (London, 1681: 106–7). Hobbs's description is clear and accurate. It is surprising, however, that it should live several days out of water.

98   Figure 8 is lost and has been reconstructed.

99   The shallowness of the English Channel was well-known from soundings and would have been elementary knowledge to one living in a harbour town such as Weymouth. Hobbs uses this information to argue for the general shallowness of the sea, taken as a fraction of the radius of the Earth. He wishes to establish that the Earth has a thin crust, to demonstrate that the crust around the poles is flexible and 'muscular' rather than rigid.

For admirable discussions of contemporary knowledge of the depth of the English Channel see A. H. W. Robinson, *Marine cartography in Britain: A history of the sea chart to 1855* (Leicester, 1962: especially pp. 28–31); and Norman J. W. Thrower, 'Edmond Halley and thematic geo-cartography' in N. J. W. Thrower and C. J. Glacken, *The terraqueous globe* (Los Angeles, 1969). Thrower points out how liberally Halley sprinkled his 1702 'Chart of the tides in the Channell' (reproduced in this edition) with fathom markings.

100   Hobbs's speculations as to the internal composition of the Earth are distinctive, but speculations on that subject were commonplace. For Halley's, see 'An account of the cause of the change of the variation of the magnetic needle: with an hypothesis of the structure of the internal parts of the earth', *Phil. trans R. Soc. Lond.* **16** 1691: 563–78, and A. Armitage, *Edmond Halley* (London, 1966: 72–4).

101   Figure 8 has been lost and reconstructed.

102   Genesis 1: 20, 21, 22:

'Let the waters bring forth abundantly fish, fowls, the moving creature that hath life, and fowl that may fly above the earth in the open firmanent of heaven. And God created great whales, and every living creature that moveth which the waters brought forth abundantly after their kind and every winged fowl after his kind and God saw that it was good.

And God blessed them, saying, Be fruitful and multiply and fill the waters in the seas, and let fowl multiply in the earth'.

103   Figures 1 and 8 have been lost and reconstructed.

104   Figures 1, 2, and 3 have been lost and reconstructed.

105   Figure 2 has been lost and reconstructed.

106   Figure 7 has been lost and reconstructed.

107   Hobbs's theory of the origin of strata on the sea bed as a result of natural processes bears some relation to Hooke's theory as set out in his *Posthumous works* (ed. R. Waller, London, 1705).

108   The Royal Exchange was built of Portland Stone. See T. F. Reddaway, *The rebuilding of London after the Great Fire* (London, 1951: 124–7, 266–77). It was designed by Edward Jarman, built 1667–9, and destroyed by fire in 1838. It is possible that Hobbs took particular notice of the Exchange while visiting the adjacent Gresham College.

109   *Cf.* the map of Weymouth and district which I have appended. Hobbs's remark about the pebbles formed from lumps of clay fallen from the cliffs is observant and accurate. The clay in question is very probably Oxford Clay.

110   Whitenore Fort most likely refers to the Nothe Fort, or The Knoll, or North Point, at the end of Weymouth Harbour, where fortifications had existed at least since Elizabethan times. It is unlikely that Hobbs is referring to the White Nose, or White Nothe, which is a head almost equidistant between Weymouth and Lulworth, and hence about seven miles away from Weymouth.

111   Genesis 1: 9: 'And God said Let the waters under the heaven be gathered together into one place, and let the dry land appear, and it was so'.

2 Peter 3: 5: 'For this they willingly are ignorant of, that by the word of God the heavens were of old, and the earth standing out of the water and in the water'.

112   A reference to Purbeck Marble.

113   2 Peter 3: 5 quoted once more.

114   For contemporary debates on spontaneous generation see note 12.

115   Genesis 1: 24.

116   About here there is a marginal note in a later hand: 'Stowes Annals'. The reference is to John Stow, *Annales of England* (London, 1592), probably referring to p. 1219. For the late seventeenth century debate on spontaneous generation, see note 12.

117   Many writers who asserted spontaneous generation believed that the Earth was more fertile for bringing forth creatures in its earlier days, and that the first species were larger than current ones. Similar beliefs are present throughout Buffon's *Histoire naturelle* (Paris, 1749–1804).

118   The worthy___ is almost certainly William Harvey, who has a passage similar to this at the end of Exercise 3 of his *Exercitationes de generatione animalium* (London, 1651). For explication *cf.* K. D. Keele, *William Harvey, the man, the physician and the scientist* (London, 1965: 192f.). It is very likely that Hobbs obtained much of his knowledge about contemporary generation theory from Dr George Garden 'A discourse concerning the modern theory of generation' which he would have read in Halley's *Miscellanea curiosa* (London, 1705, I: 142–54).

119   Hans in Kelder: Dutch for 'Jack in the cellar', *i.e.*, colloquial for an unborn child. Hobbs is of course utterly opposed to preformationism. His discussion of doctrines of generation seems to echo Dr G. Garden, 'A discourse concerning the modern theory of generation', reprinted in Halley's *Miscellanea curiosa* (London, 1705, I: 142–54).

120   *Cf.* P. C. Ritterbush, *Overtures to biology* (New Haven and London, 1964: 88–99), and N. E. Nordenskiöld, *The history of biology* (New York, 1929: 197).

121   Leffas: not identified. For the debate about the nutrition of plants and spontaneous generation see J. Woodward, 'Some thoughts and experiments concerning vegetation', which Hobbs would have read in Halley's *Miscellanea curiosa* (London, 1705, I: 205–44).

122   Hobbs had evidently not seen the letter on this subject to William Cole printed at the end of Robert Hooke's *Lampas, or description of some mechanical improvements of lamps and waterpoises* (London, 1677). Cole's letter was entitled 'An observation about the seed of moss'. His reference to ferns is pp. 47–48.

123   Figures 1 and 8 are missing and have been reconstructed.

124   One of many indications that Hobbs does not accept the short time-scale entailed by a literal reading of the Bible.

125   An obvious attack on Thomas Burnet's *Telluris theoria sacra* (London, 1681 and 1689; English translation, 1684 and 1690). For the problem of the function of mountains in the system of nature as understood in the late seventeenth century see G. L. Davies, *The earth in decay* (London, 1969: chs i and ii); M. H. Nicolson, *Mountain gloom and mountain glory* (Ithaca, New York, 1959: chs i–iii). Yi-fu Tuan, *The hydrologic cycle and the wisdom of God* (Toronto, 1968).

126   For the great comet of 1680, see C. A. Ronan, *Edmond Halley: genius in eclipse* (London, 1969: 59f.). For Halley's views of comets as possible agents of geological destruction, see E. F. MacPike, *The correspondence and papers of Edmond Halley* (London, 1932: 234), and for W. Whiston's similar ideas, see his *New theory of the earth* (London, 1696).

127   Malachi 4: i: 'For behold, the day cometh, that shall burn as an oven'.

128   For contemporary debate about the likely end of the earth see E. L. Tuveson, *Millennium and utopia* (Berkeley, 1949).

129   Note again the implication of a high time scale.

130   For contemporary debate on the problem of the origins of rivers see A. K. Biswas, *History of hydrology* (Amsterdam, 1972); G. L. Davies, *The earth in decay* (London, 1969: chs i–ii; and for a contemporary account see above all R. Plot's *De origine fontium* (Oxford, 1685).

131   For contemporary debate on the problem of the age of the Earth see S. E. Toulmin and J. Goodfield, *The discovery of time* (Harmondsworth, 1967: chs iii and iv); H. Meyer, *The age of* *the world* (Allentown, Pa, 1951); F. Haber, *The age of the world: Moses to Darwin* (Baltimore, 1959: chs i–iii).

132 2 Peter 3: 8.

133 Genesis 1: 16: 'And God made the two great lights; the greater light to rule the day, and the lesser light to rule the night; he made the stars also'.

134 Hooke had employed similar arguments from fossils to press for a high antiquity for the Earth. See his *Posthumous works* (ed. R. Waller, London, 1705).

135 Figure 3 is missing and has been reconstructed.

136 Shaston is Shaftesbury. In fact the trees were presumably not petrified, but rather dated from historical times. For although fragments of lignite would be found in the Kimmeridge Clay near Shaftesbury (which is often carbonaceous), trunks of trees of the size quoted by Hobbs would not be found in the rock beds themselves. Joshua Childrey and John Aubrey had visited in 1668 a hill near Shaftesbury which contained plenty of Cocklestones. See A. R. and M. B. Hall, eds, *The correspondence of Henry Oldenburg* (Madison, 1968, **5**: 456).

137 It is not clear whether Hobbs had visited the repository of Gresham College (*i.e.*, the Royal Society), or had merely seen Nehemiah Grew's *Musaeum Regalis Societatis* (London, 1681) where petrified wood is mentioned on pp. 265–74.

138 For background to the problem of the Deluge, see D. C. Allen, 'The legend of Noah', *University of Illinois studies in language and literature* **33** 1949. Hobbs's almost total dismissal of discussion of the Deluge is most unusual for its time.

139 Robert Plot's discussion of Fairy Rings is in his *Natural history of Stafford-shire* (Oxford, 1686: 9–20). John Aubrey's *Natural history of Wiltshire* (ed. J. Britton, London, 1857: 37) contains a similar discussion.

140 The chief theorist of the view that all rocks were dissolved at the Deluge was John Woodward in his *An essay towards a natural history of the earth* (London, 1695).

141 Genesis 8: 4: 'And the Ark rested in the seventh month on the seventeenth day of the month, upon the mountains of Ararat'.

142 This is a slightly surprising statement, for till now Hobbs has expressly been confining himself to what is *naturally* possible. Now he seems also to rule out the possibility that the Deluge could ever miraculously have had the effects which Woodward and others supposed.

143 Psalm 14: i. 'The fool hath said in his heart, There is no God. They are corrupt, they have done abominable works, there is none that doeth good'.

144 For the extensive contemporary argument over the supposed rise of 'atheism' *cf.* R. L. Colie, *Light and Enlightenment* (Cambridge, 1957); M. C. Jacob, *The Newtonians and the English Revolution, 1689–1720* (Hassocks, Sussex, 1976); J. Redwood, *Reason, ridicule and religion* (London, 1976).

145 For discussion of the possible date of composition of the Postcript, see p.31.

146 A very large number of maritime manuals were produced in Dutch with the title *De Lichtende Colomne ofte Zee-Spiegel* throughout the seventeenth century. The first I have been able to trace in England (*The lightning colomne or sea-mirrour*) was printed in 1654 at Amsterdam (there is a copy at the National Maritime Museum at Greenwich: I am particularly grateful to Miss Margaret Deacon for her help in locating and describing this). Thereafter *Lightning columns* appeared regularly in English, every few years, and it is perhaps surprising that Hobbs had not seen one earlier. For a listing of English *Lightning columns* see the definitive work of I. C. Koeman, *Atlantes neerlandici* (Amsterdam, 1970, **4**: xxv–xxxviii). Koeman has not been able to trace, however, a copy surviving of the 1701 edition. The title page of a 1692 edition is eloquent of the contents:

> The / Lightning Columne, / or / Sea-Mirrour, / contaigning the / Sea Coasts of the Northern and Eastern / Navigation: Setting forth in divers necessaire Sea-Cards, all the Ports, / rivers, bayes, roads, depths and sands, very curiously placed on its / due Polus height furnished. With the discoveris of the chief countries / and on what cours and distance they lay one from another. Never / theretofore so clearly laid open, and here and there very diligently / bettered and augmented for the use of all seamen. / As alsoo the / situation of the northernly countries, as islands, the Strate Davids, the isle of Jan Mayen, Bears Island, / old Greenland, Spitsbergen and Nova Zembla: adorneth with many sea-cards and discoveries. / Where unto is added a brief instruction of the art of navigation, together with / new tables of the suns declination, with an new almanach. / Gathered out of the experience and practice of divers pilots and lovers of the famous art of navigation. / At Amsterdam, Printed by Casparus Loots-man, Bookseller in the Loots-man, upon the Water. Anno 1692. With Privilege for Fiftheen Jears. (see Koeman **4**: 254).

For a good discussion of the evolution of the sea atlas see A. H. W. Robinson, *Marine cartography in Britain* (Leicester, 1962: ch. ii).

147 Hobbs's map of the world is, unfortunately, lost, and his descriptions of it hardly enable one to make a hypothetical reconstruction. Presumably it looked something like a hybrid between Halley's map of the trade-winds and his 'Chart of the tides in the Channel'. For the later development of the attempt to plot the times of tides across the globe, and to understand their regularities and laws, *cf.* W. Whewell, *Essay towards a first approximation to a map of cotidal lines* (London, 1833)—see pp. 226–7 for the irregularities of the tides at Weymouth.

148 Herman Moll (d. 1732) was a Dutch geographer, working mainly in Britain, who engraved maps for a large number of early eighteenth century geographical works, above all *The compleat geographer*, 3rd ed. (London, 1709), and *A view of the Coasts, Countries and Islands within the limits of the South Sea Company* (London, 1711).

149 John Seller was the author of several works dealing with the coastal geography of Britain, above all the *Atlas maritimus* (London, 1675) and *The coasting pilot* (London, 1673). See A. H. W. Robinson, *Marine cartography in England* (Leicester, 1962: 38–42 and Appendix J).

150 John Thornton was a geographical and navigational author. His works include the *Atlas maritimus novus* (London, 1708); and *A correct table of the latitude and longitude of the most noteable capes* (London, 1706).

151 William Dampier (1652–1715), bucaneer, pirate, navy captain and circumnavigator, and the most popular writer of naval voyages of his age; author in particular of *A new voyage round the world* (London, 1697).

152 Woodes Rogers (d. 1732), sea captain and governor of the Bahamas. Author above all of *A cruising voyage around the world* (London, 1712).

153 This is of course not true: the motions of the tides are infinitely more complex, and still today hardly understood.

154 See Captain William Dampier, *Voyages and descriptions in three parts. Pt. iii. A discourse of trade winds, breezes, storms, seasons of the year, tides and currents of the Torrid Zone throughout the world; with an account of Natal in Africk, its products, negro's etc.* (London, 1699: 90ff.).

155 *Ibid.*

156 Halley's 'maps' referred to here would be his 'Chart of the tides in the Channell' (London, 1702); 'An historical account of the trade winds and monsoons', *Phil. trans R. Soc. Lond.* **16** 1686: 153–68; his 'Isogonic Map of the Atlantic' (London, 1701); his 'Isogonic Map of the World' (London, 1702). There is an excellent discussion of the theoretical importance of Halley's maps in Norman J. W. Thrower, 'Edmond Halley and thematic geo-cartography', in N. J. W. Thrower and C. Glacken, *The terraqueous globe* (Los Angeles, 1969), which also contains an admirable bibliography on Halley. Halley's chart of the English channel is reproduced and discussed in Derek Howse and Michael Sanderson, *The sea chart* (Newton Abbot, 1972: 80–1).

157 Figure III, Plate 3, is lost, and I have not felt confident to reconstruct it.

158 Figures I and II, Plate 3, are lost and I have not been able to reconstruct them.

159 This is necessarily rather obscure in the absence of Hobbs's maps.

160 See note 157.

161 See note 156.

162 The tides around Weymouth are notoriously complex, partly due to Chesil Bank.

163 See note 157.

164 See note 154.

165 In the Gulf of Mexico.

166 San Miguel in the Gulf of Panama.

167 In the Cape Verde Islands.

168 See notes 157 and 158.

FIGURE 1   Mentioned in Hobbs's manuscript on page 1 (in this edition: page 39), 15 (48)
16 (49), 91 (97), and 111 (107).

   The original diagram is lost and has here been reconstructed. It does not seem to refer to
any specific location. Its chief aim is to demonstrate that strata are found parallel to each other
and generally parallel to the horizon. In connection with this diagram, Hobbs mentions
(MS 91; this edition, 97) that fossils of 'Oysters, Wrincles, Cockles &c' are found within the
strata.

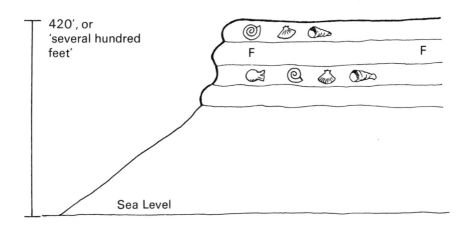

FIGURE 2 Mentioned in Hobbs's Manuscript on pages 2 (this edition: 39), 16 (49), 17 (50), 91 (97), and 93 (97).

The diagram is lost and has been reconstructed. Hobbs is expressly referring to the cliff scenery of the Isle of Portland, 'soe many hundred feet higher than the Ocean', his most precise estimate being 420'. Hobbs notes that the strata to which he is referring extend downwards from the surface for between 60' and 100'. He is clearly describing Portland stone, and points out that it is highly fossiliferous. It is underlain by Portland sand and Kimmeridge clay, neither of which is well exposed—partly because rubble from the quarries was doubtless obscuring the lower strata. Landslips caused by the failure of the lower soft beds, thereby bringing down blocks of Portland stone, were also significant in this effect. The barren bed F.F.F. might refer to a number of beds within the Portland stone such as the Whit bed (7–15' thick) which is a limestone largely composed of tiny shell fragments that might have escaped notice. Identification of the bed F.F.F. as the Whit bed is not certain, and there are other possibilities, though the Whit is considered an excellent freestone and Hobbs would probably have had many opportunities for observing it.

At the depth of 60–100' which Hobbs mentions, he is undoubtedly referring mostly to the Portland stone, but as the Purbeck beds overlie the Portland stone on the Bill, there is little doubt that Hobbs included them in his estimate.

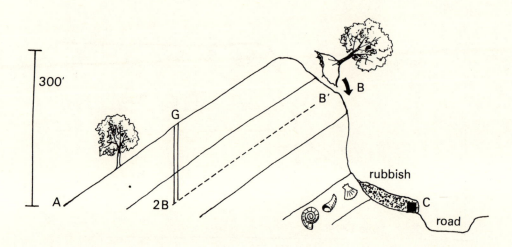

FIGURE 3   Mentioned in Hobbs's Manuscript on pages 1 (39), 15 (48), 16 (49), 17 (50), 18 (50), 91 (97), 119 (112).

This diagram is lost and has been reconstructed. It is designed to show that strata are generally found parallel to each other, and in hilly country parallel to the line of the hills. Hobbs notes how strata run unbroken, at an angle of perhaps 20 or 40° along one side of the hill (*i.e.*, dip slope) but are broken off sharply along the other side (*i.e.* scarp face). He notes how materials (including trees) become detached from near the top of the scarp face and tumble down to form mounds of rubbish.

Presumably this diagram illustrates his discussion on p. 120 (112) of finding tree trunks having tumbled down from a high hill near Shaston (= Shaftesbury). In the Shaftesbury area there are several beds which Hobbs might have had in mind when referring to his hill. The detachment of bed A suggests that it was a competent (geological speaking) bed such as limestone underlain by a softer sand or clay. There seems to be no definite way of discovering which hill and strata Hobbs had in mind.

In his discussion of how a knowledge of the geometry of stratification can help discovery of mines, etc., Hobbs argues that if a seam is found at 2B, 130' down from G, one can predict that the same seam will be found at B1, three or four feet down from B (by which he presumably means under the top-soil).

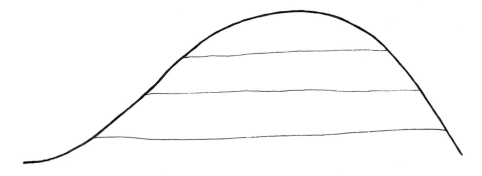

FIGURE 4 Mentioned in Hobbs's manuscript on pages 1 (39), 16 (49), 93 (98). This diagram has been lost and is here reconstructed.

Hobbs uses this diagram to illustrate his contention that strata are basically found parallel to each other and to the horizon, and that this can be the case even with land considerably elevated above sea level. I assume that he particularly has in mind the topography of cliffs. Hobbs does not make it clear what relation he imagines the elevated land to bear to the surrounding terrain. Whether Hobbs considered that the elevated portion achieved its position by the action of faults or the removal of surrounding rocks is uncertain.

FIGURE 5   Mentioned in Hobbs's manuscript on page 16 (49). This diagram has been lost and is here reconstructed.

In it Hobbs embodies his denial that in hilly country the strata ever run completely parallel to the three-dimensional topography. He argues that the structure of hills is always as found in diagrams 3 and 4. This serves to illustrate his conviction that hills have been created by elevation from below, rather than by any kind of precipitation from above. Elevation from below, Hobbs believes, must necessarily result in upland areas possessing at least one jagged end and broken face (as in a cliff).

Hobbs's denial of this possibility shows that he had no real awareness of the anticlines of Dorset.

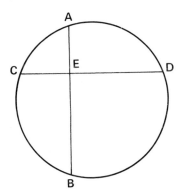

FIGURE 6 Mentioned in Hobbs's manuscript on page 25 (55). This diagram has been lost and is here reconstructed.

It illustrates Euclid li. 3, pro. 35. For explanation see footnote 55.

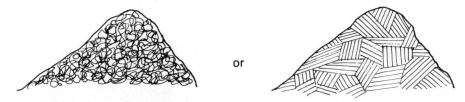

FIGURE 7 Mentioned in Hobbs's manuscript on pages 16 (49) and 92 (97). This diagram is lost and has been reconstructed.

It is intended by Hobbs to demonstrate an impossibility in Nature—*i.e.*, the strata found in no order whatever. It is not clear whether Hobbs had in mind a situation in which small particles of different rock types would be accumulated chaotically, or whether he was thinking of small wedges of strata abutting against each other at random angles. His aim is to show that strata are not the product of any kind of chemical precipitation, or the remains of the Deluge, or the remnant of some other disordering force in Nature, but rather have been gradually formed on the sea-bed, and subsequently regularly and gently uplifted.

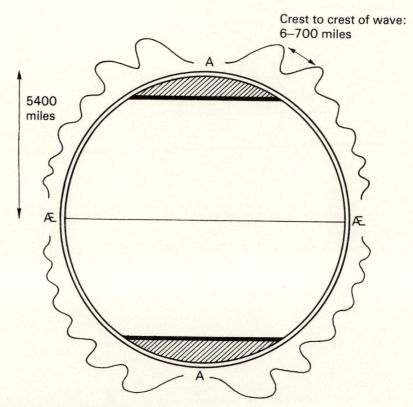

Crest to crest of wave:
6–700 miles

5400
miles

FIGURE 8 Mentioned in Hobbs's manuscript on pages 2 (39), 81 (91), 83 (92), 91 (97), 111 (107). This diagram has been lost and is here reconstructed.

In his text Hobbs refers both to a Figure 8 and to a Figure VIII. I have assumed that he had in mind the same diagram, and have here combined the information which relates to both of them. Hobbs is chiefly concerned in this figure to demonstrate two points:

(*a*) that the shell, or crust, of the globe is very thin indeed as a fraction of the diameter of the globe itself. His estimate is the—reasonably accurate—one of ten or twelve miles, which, as he says, can only properly be represented on his diagram by the thickness of the stroke of a pen. He is concerned to show that strata within this shell are horizontal. He infers the thinness of the shell from the shallowness of the sea in the English Channel, and by analogy with the *orbis minor*.

(*b*) that at the two poles of the globe are to be found musculous, cordious areas, whose pulse creates the motion of the tides.

On pages 134–8 of the manuscript (in this edition, pages 120–123) Hobbs discusses the wave motion by which this pulse of the tides is communicated from the poles to the equator (being conveyed some 600–700 miles per tide, and diminishing towards the equator). He seems to have illustrated this by his map, which I have not endeavoured to reconstruct. I have however inserted an indication of such a wave motion emanating from the poles as an indication of what Hobbs may have had in mind.

PLATES

Please note that Plate I is a facsimile of Hobb's Title Page and is to be found on page 36 opposite the transcription.

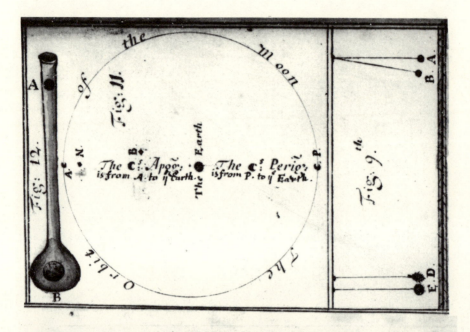

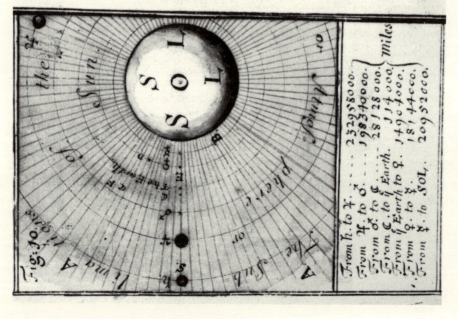

PLATE II This plate reproduces the one surviving page of figures drawn by Hobbs himself (figures 9, 10, 11 and 12). It exists as a loose sheet in the treatise. The meaning of the figures is fully explained by Hobbs in his text.

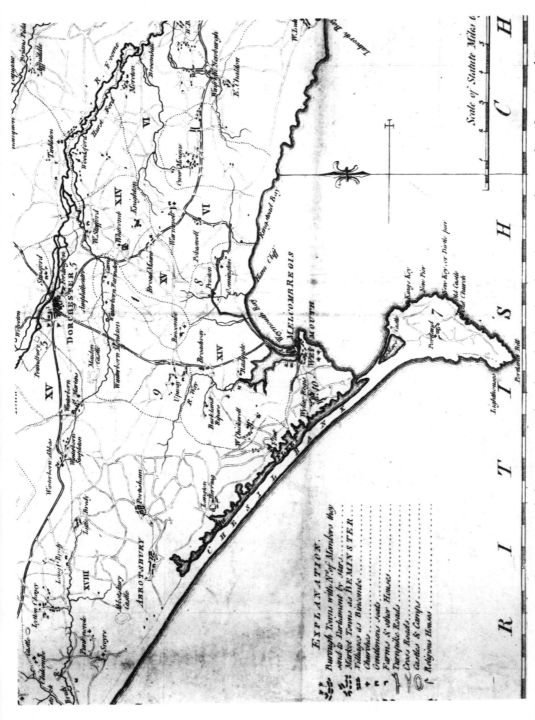

PLATE III Part of South Dorset, including Weymouth and Portland. Taken from 'A Map of Dorsetshire from Actual Surveys and Records of the Country' by J. Bayly, 1773 (also printed in Hutchins's *History of Dorset* (1774)). Its scale is 69½ statute miles to a degree.

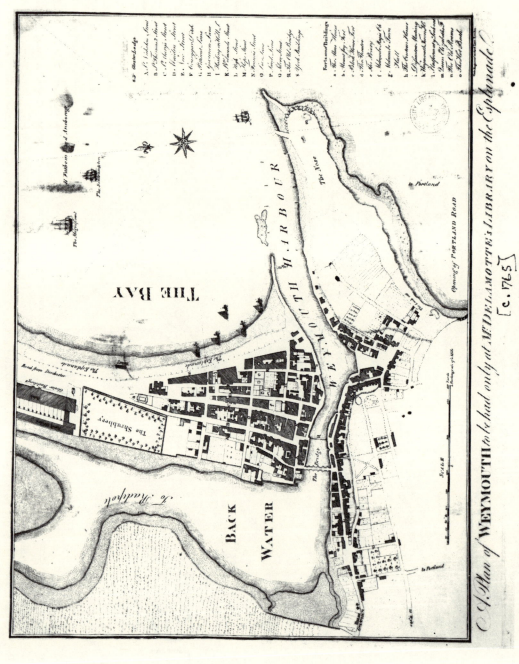

PLATE IV   Weymouth c. 1765. Taken from 'A Plan of WEYMOUTH to be had only at Mr DELAMOTH's LIBRARY on the Esplanade'. About this time Weymouth was first beginning to develop as a resort.

PLATE V 'A New and Correct Chart of the Channel between England and France with considerable Improvements not extant in any Draughts hitherto publish'd, shewing the Sands, Shaols, depths of Water and Anchorage, with ye flowing of the Tydes, and setting of the Current as observ'd by the learned Dr Halley: sold by Mount and Page on Tower Hill' [1702].

Halley's Chart of the Channel was undoubtedly much used by Hobbs, for it recorded both the times of the tides, and also the depths of the Channel. The legend reads:

In this Channell Draught, the smaller Figures are the Depth in Fathoms, The Litteral or Roman Figures shew ye Hour of High-Water, or rather ye End of the Stream that setts to ye Eastward on ye Day of ye New and Full Moon. Add therefore ye time of the Moons Southing or Northing to y^e Number found near y^e place where yor Ship is, & y^e Sum shall show you how long y^e Tide will run to y^e Eastward. But if it be more than 12 subtract 12 therefrom. The Direction of y^e Darts shew upon what Point of y^e Compass y^e Strength of y^e Tide sets = All Masters of Ships, and others, who shall have opportunity to observe y^e Depths, with Certainty in respect of y^e Place, are desired to communicate them to y^e Publisher hereof.

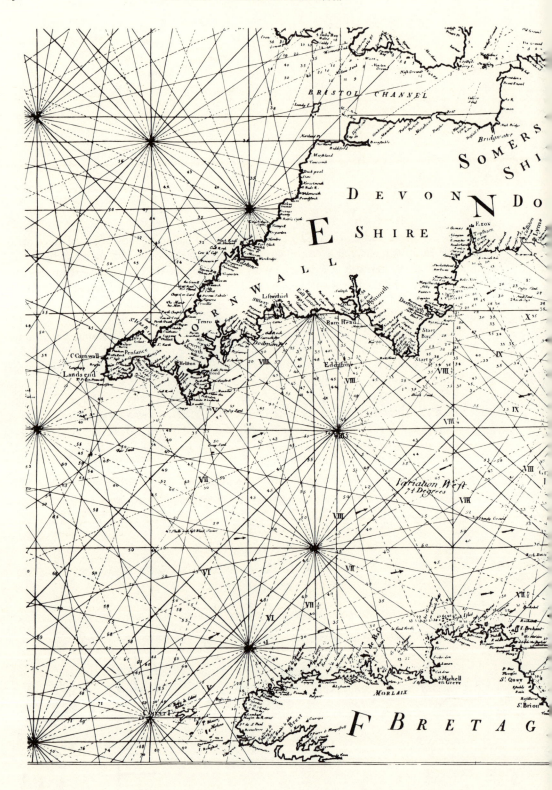

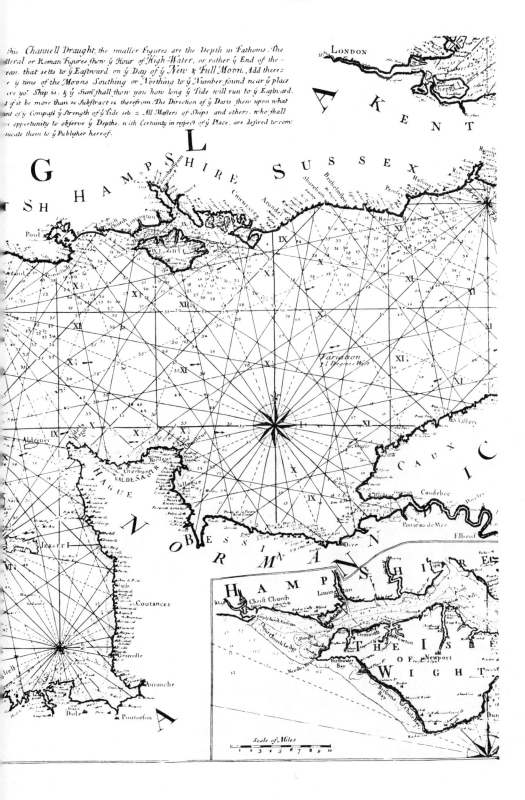

this Channell Draught, the smaller Figures are the Depth in Fathoms. The lateral or Roman Figures shew y^e Hour of High-Water, or rather y^e End of the stream that setts to y^e Eastward on y^e Day of y^e New & Full Moon. Add there-to ye time of the Moons Southing or Northing to y^e Number found near y^e place where yor Ship is, & y^e Sum shall shew you how long y^e Tide will run to y^e Eastward. but if it be more than 12 Substract 12 therefrom. The Direction of y^e Darts shew upon what point of y^e Compass y^e Strength of y^e Tide sets = All Masters of Ships and others, who shall have opportunity to observe y^e Depths, with Certainty in respect of y^e Place, are desired to com-municate them to y^e Publisher hereof.

LONDON

KENT

GL

HAMPSHIRE SUSSEX

NORMANDIE

HAGUE

CAUX

Scale of Miles

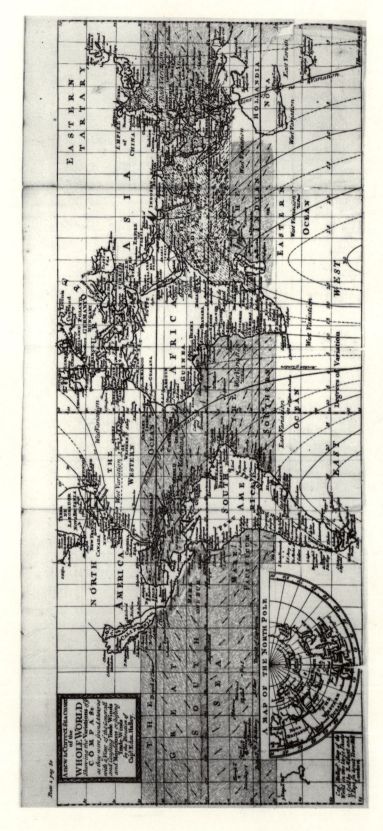

PLATE VI Edmond Halley's 'A New Correct Sea Chart of the WHOLE WORLD, shewing the Variations of the COMPASS, as they were found ano 1700, with a View of the Generall and Coasting Trade Winds and Monsoons or Shifting Trade Winds by the Direction of Capᵗ Edm. Halley', here reproduced from Halley's *Miscellanea curiosa* (London, 1705, I: facing p. 81).

Hobbs made extensive use of this map. I assume that his lost world map of the times and directions of the tides incorporated features of this map, together with some taken from Halley's *Chart of the Channel*.

LIBRARY OF DAVIDSON COLLEGE

Books on regular loan may be checked out for **two weeks**. Books must be presented at the Circulation Desk in order to be renewed.

A fine is charged after date due.

Special books are subject to special regulations at the discretion of the library staff.

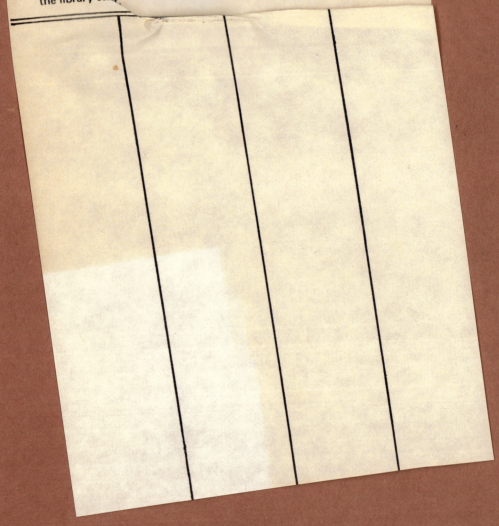